D0025023

RELENTLESS IMPROVEMENT

TRUE STORIES OF LEAN TRANSFORMATIONS

RELENTLESS IMPROVEMENT

TRUE STORIES OF LEAN TRANSFORMATIONS

Bill Trudell

CRC Press is an imprint of the
Taylor & Francis Group, an **informa** business

A PRODUCTIVITY PRESS BOOK

CRC Press
Taylor & Francis Group
6000 Broken Sound Parkway NW, Suite 300
Boca Raton, FL 33487-2742

Printed in the United States of America on acid-free paper
Version Date: 20121030

International Standard Book Number: 978-1-4665-5430-6 (Hardback)

Library of Congress Cataloging-in-Publication Data

Trudell, Bill.
Relentless improvement : true stories of lean transformations / Bill Trudell.
p. cm.
Includes bibliographical references and index.
ISBN 978-1-4665-5430-6 (hardcover : alk. paper)
1. Lean manufacturing. 2. Manufacturing processes. 3. Production management. 4. Organizational change. 5. Industrial efficiency. I. Title.

TS155.T79 2013
658.5--dc23 2012032353

Visit the Taylor & Francis Web site at
http://www.taylorandfrancis.com

and the CRC Press Web site at
http://www.crcpress.com

Dedication

As a father, I have been blessed at the highest level.

I dedicate this book to my four children

Leslie, Adam, Cayman, and Harrison. Being your father

is and has been an incredible experience, and I am as

proud as I can be of each and every one of you.

—Dad

Contents

Introduction

I am a "manufacturing guy" at heart. I have worn out at least a hundred pairs of work boots in my career. My Dad and my brother are "manufacturing guys." My grandfathers, my uncles, and a lot of my cousins...they are "manufacturing guys." My manufacturing career started the week after high school when I walked into a factory to start work. What a world...screw guns, impact wrenches, spray guns, staplers, hammers, chisels, wrenches, crimpers, cutters, brushes, mixers, sandpaper, sanders, utility knives... and people, which was the best part. I have now been associated with factories in some sort or fashion for most of the past thirty years. During that time I have become a passionate scholar of manufacturing. Over the years I have spent a great deal of time studying the best ways to achieve success in manufacturing. Often, I was surrounded by situations that could have been resolved by applying them, yet management was not committed, or didn't get it. Sometimes frustrating, but I always found a way to get at least some of the practices going. There were success stories, too.

My manufacturing journey started in 1975 and transitioned to my Lean manufacturing journey starting in 1990 after reading *Just in Time for America* by Henry Wantuck. What a great book for its time, and actually it still is today. I can see the book on my bookshelf with a quick glance as I write this. Having a career in manufacturing has been unbelievably rewarding and very challenging, unbelievably challenging, but fun. Twenty-five years in manufacturing, and a lot of devout studying and hands-on experience have developed my perspective.

As I look back at my lifetime, I see transitions not only in myself, but also the transitions in American industry. When I started work in a factory in 1975, Toyotas were funny-looking foreign cars and anything bought in Japan was cheap. Having "Made in Japan" on it was not a good thing, and usually meant putting it back on the shelf. I never would have guessed that years later I would be a devout scholar of Japanese manufacturing techniques. Little did I know while growing up outside of Detroit where the Big Three auto giants ruled, that forty years later it would be necessary for the U.S. Government to bail them out, while Toyota and the other Japanese auto companies continued to stand on their own.

Reflecting on my manufacturing experience, I see that in the 1970s some companies were following practices that were parallel with Lean principles. They just were not called Lean because the vernacular was not there, as Lean had not been integrated into the U.S. economy yet. They were also just practices as opposed to an all-encompassing Lean culture. These approaches were just common-sense things to do, like focusing on high quality or doing things right the first time. The focus was clearly on getting product out the door. At that time, quality was one-dimensional, as the only benefit companies saw in it was not upsetting the customer with cosmetic or functional issues. The fact that the product was produced in an environment ripe with rework and do-overs was not considered. The significant amount of money that could be delivered to the bottom line via eliminating rework and warranty was not really showing up on company radar screens. The market was so good then that they could be wasteful and still make money! Some companies did manage work-in-process, and some had work procedures or standard work, but they were deployed more from an organizational perspective as opposed to an element of an overall philosophy for running a business. Some companies clearly listened to their customers and others did not. They were deciding what their customers wanted instead of asking them, and then some companies guessed well.

My experience in the 1980s was that companies were trying to apply process disciplines to maintain order in their day-to-day operations. They really did not understand the important role of the length of product cycle times. Salespeople were paid and given very large bonuses to sell, sell, sell, and manufacturing staffs were paid to ship, ship, ship. In many places, there was more emphasis on in-process quality because it interfered with shipping product. The "customer" was very seldom talked about. It seemed that everyone took for granted that there were always a lot of customers. If you built it, you could sell it. Management was the elite. They got paid to think and workers got paid to work. If management wanted to know something, they would ask—otherwise keep working.

The 1970s ended with a recession, and so did the 1980s. Business picked up again in the 1990s, but American manufacturers were becoming like the proverbial frog put in a pot of cool water sitting on top of a hot stove. They were on their way to boiling to death and did not have a clue what was going on. All of a sudden, customers were becoming demanding; markets were changing, shrinking, or disappearing. Competition was getting greater and greater in all industries. Being profitable required managing cash flow, managing working capital, managing inventories. Wasteful processes were becoming unacceptable, and management itself did not have all the answers that most likely existed in the minds of their employees.

In the late 1990s, success stories of Japanese manufacturing techniques were becoming more common. JIT (Just In Time) was the buzzword. Right away, management saw JIT as having low inventories as opposed to being one of many benefits of a philosophy for running a business. They really did not get that in order to realize the benefit of being able to operate with the proverbial low inventories of JIT, numerous disciplines throughout the business had to be implemented. In fact, if you asked most people back then what JIT was, most would tell you it was running a business with low inventories and that it was a purchasing function. In fact, some companies were so shallow as to just lower their inventories and claim they were JIT. Yet, they did not arrive at low inventories, they just went there, absent of developing the disciplines required to operate at lower inventories. Then when they could not operate at lower inventory levels, they claimed that JIT did not work.

Comments were being made that the Japanese were going to put the American automotive companies out of business. There were a lot of people who believed this. The Japanese can be credited with putting some fear in the Big Three and inspiring a greater focus on quality. U.S. automakers had no choice. Honda, Toyota, Subaru, and Nissan were all making cars that had many fewer quality defects. I personally bought a new Honda car in 1982 and a new Toyota pickup truck in 1994. I can honestly say that I never had to take either one of them into the dealer to have something fixed. I drove them, maintained them, and sold them at a good resale price. I did find a Kanban card under the seat in my truck one day. I got a kick out of it. You know what was cool? I used it as a bookmark in my *Just in Time for America* book. I just got up and checked the book on my shelf, and guess what? I still have it. In fairness, I have a Big Three truck today and am very pleased with it (other than the dings and dents from my teenage daughter learning to drive).

My hunger for knowledge of all things manufacturing was growing. I could not get enough of reading books, learning, touring factories—you name it. I was pursuing the APICS (American Production and Inventory Control Society) certification in the early 1990s. The certification required passing five rigorous examinations, one of which covered JIT manufacturing. To prepare, I purchased JIT books and signed up for a two- or three-day review course in the suburbs of Chicago. The reason I point that out is that in the 1990s, the quality of APICS training in the Chicago area was incredible. The base of large companies provided for some very talented instructors. I remember thinking that some of these people were more knowledgeable than some of my college professors. In fact, it was not unusual to have a Ph.D. teaching a review course. I was so fortunate to be working in that area.

Two guys teamed up to lead the JIT review course. They were incredibly knowledgeable, had great senses of humor, and kept the group engaged every minute of every day during the training. They were fantastic. I am a problem solver and a strategist at heart. I like to fix things like factories, processes, departments, and I am big on developing a strategy at the beginning of every effort. I plan the work, and work the plan. The tenets of JIT or Lean caught my attention immediately. They offered solutions, strategies, and rhyme and reason to tackling the manufacturing environment. They offered such great things to focus on…cycle time reduction, set-up reduction, quality at the source, one-piece flow, mistake proofing, and on visual factories, voice of the customer, house of quality, and on and on. Furthermore, they tied it all together into a philosophy for running a business. This is where my Lean journey officially began.

Throughout my career, I have been passionate about Lean and, actually, I looked at the jobs I had as just opportunities to apply and learn more about Lean. As I started reading about Lean, I remember thinking how much sense it made, and furthermore, it was just plain interesting. The more I read, the more it just made sense. Honestly, I am a bit of a "right-brainer," and I had to re-read some paragraphs a few times. Taking it a step further, it took a lot of Lean books, articles, training, and experience to really get Lean to "click" in my mind. I think it is the most complicated simple thing I have ever encountered! To this day, I continue to enjoy cracking the cover of a new book on some Lean or Lean Six Sigma topic. I have two books on my nightstand as I write. I learn something new from every book, and from every week that passes. At fifty-five years old, I still find myself hearing the clicks and saying to myself, "Oh, OK, I get it now."

For more than twenty years, I have studied Lean, taken Lean-related training, and applied the tools. At the same time, I have never been part of a company that really embraced Lean from the top. I have been fortunate to work for company presidents who generally supported the use of Lean tools and approaches, but never the "whole ball of wax." Like a lot of people, there were times I really wanted to take Lean to a much higher level, or see it more widely dispersed throughout the company, but the culture and the top would not support it. There is a big difference between tolerating

Lean and fully supporting it. Being a positive and can-do person, I have always focused on what I could do, not what I could not do. So, because I could not do Lean all the way, I did Lean things, and a lot of them. This book is about my relentless efforts to improve during my career, and doing so using the tools of Lean Six Sigma. I call it "relentless improvement" because improvement must be relentless. You can never let up, ever.

I have mapped value streams, created flowcharts, conducted time studies, created standard work, calculated Takt times, level-loaded work centers, taught quality at the source, asked why five times, done fishbone analysis, flown to Japan and reviewed Japanese factories that practiced Lean under the guise of a quality audit, participated in design for manufacturing and assembly (DFMA), led set-up reduction activities, applied Kanban, and led 5S implementations. But I never worked for a company with a top-down commitment to applying Lean throughout the company. So I did Lean things, projects, value-stream or process mapping events, Kaizen events, and listened to operators, and always solicited their input.

Whether or not you are in a company that supports Lean from the top, you can still get some great results by using Lean tools to accomplish improvement projects. You do not have to be in a Lean organization to tap into the knowledge of operators to root cause and solve process problems. I will review the many instances where I have applied the principles of Lean in managing projects over the years. I have had many small successes by applying Lean, and some notable ones too. One of my passions has always been to turn chaos into order. I have had the experiences of leading two factory turnarounds, two factory start-ups, and have led or been involved in forty-seven new product introductions. Some of these introductions were horrendous. We will discuss that later. And remember when the wife of the plant manager in *The Goal* left him and then came back? Well, mine never did!

I consider myself fortunate to have spent the majority of my career manufacturing products that are great fun—boats. I loved boats when I was a kid. My first boat was purchased for $5 from an old man, Mr. Decou, our next-door neighbor in Michigan. He was an old walleye fisherman and was famous for making "Decou McGintys" in his basement. A McGinty was a tiny fishing lure used in trolling for walleye…kind of a colored, cone-shaped thing with a propeller on one end. They were sought after by fishermen all over the state of Michigan. After getting that first boat, I was hooked. As a young kid, a boat was a real sense of freedom and adventure. I kept the boat by a dock in a small river. My friends and I would pack sandwiches, grab our fishing rods and BB guns, and head out for the day. Because it was a wooden boat, each trip started by bailing it out with a Clorox® jug made into a bucket or a hand bilge pump. Many times, there would be only three or four inches of the boat sticking out of the water.

That love of boats turned into a passion for building them, which turned into a passion for manufacturing and Lean manufacturing. So all the Lean projects, efforts, and exciting Lean things I am going to write about take place in a unique industry, the marine industry. It is my hope that you will enjoy the background while I write about applying Lean. I mean, hey, I could have spent my career making screwdrivers or oven mitts.

During my career, I have met and worked with highly talented and highly educated people, and have made friendships that will last a lifetime. I have had the pleasure of meeting many unbelievably inspiring leaders, and learned a lot from them. On the other hand, I have also worked with some "not so good" leaders, highly political self-serving executives, and incompetent and inept people. My perspective during my career has been to try to have fun and achieve excellence; to get things right; to achieve as close to perfection as possible in processes. The Japanese term for playground (*Gemba*) works for me because, to this day, a factory is still that…a playground. I still get excited when my shoes hit a factory floor and the operators, employees, associates, whatever you want to call them, bring a smile to my face. They truly do have all the answers if you have the right questions. To me, one of the biggest benefits of Lean is that it teaches you what questions to ask. How about "why?" Now there is a powerful question.

I have always tried to stay current with the cutting-edge approaches and philosophies and have worked to apply them. These approaches were tried and proven, and in use in many high-profile companies. Even with their obvious success, many leaders I worked with or for were dismissive of them. These same leaders would stress the need to resolve serious issues in the company, yet dismiss these new and proven approaches. It was frustrating to be in situations where upper management was pounding on cost issues, issues around getting product out, out of control warranty, historical problems that had been around for years, all things that Lean addresses in a huge way. Often, when the idea of embracing Lean principles or approaches was brought up, eyes would roll or the suggestions were quickly dismissed and the subject was changed. Improvement was many times a series of simultaneous nonintegrated actions absent an overall comprehensive goal or strategy.

The times that I was able to apply the new knowledge and approaches found in Lean I enjoyed greatly. I did have bosses and situations that gave me a lot of leeway. To me, it was a blast to work with employees reaching their full potential and see them improve their processes and work quality. To this day, one of my favorite things to hear when walking through a factory is an employee shouting out, "Hey, Bill," asking me for help. Although it still happens in my role as a Lean leader, it is one of the things I miss most about being a factory manager. An employee once told me that another employee told him to bring his problem to me because "that man gets stuff done." That is music to any leader's ears.

So, the Toyota production system or Lean manufacturing over time, more and more became the pinnacle of reference for me in all things manufacturing. Learning and applying these principles have been the most rewarding parts of my career. To work with operators or employees in teaching them to use their knowledge to solve problems or improve their processes has created many meaningful experiences for me. I get such a kick out of working with everyday employees to generate fantastic solutions to problems. It is also fun to see them bring their families through their departments during company open houses and watch them proudly point out all the improvements they made.

The use of Lean principles is seriously important in my eyes, because effectively creating a Lean culture and applying it ensure the highest probability of success any company can achieve. If anything can add to job security, it is Lean. After all, Lean is about giving customers the optimal value at a cost they are willing to pay. That is a big deal, and what can be more important than pleasing customers and controlling costs? Manufacturing is a tough business, and it is nice to have a set of good tools…good stuff.

So, with all that said, in this book I am going to take the reader through my career experiences in manufacturing and outline the transition that brought us to where we are today based upon my perspective. I will start in 1975 when I first went to work in a factory and go all the way through to today. It is my hope that this book will do two things. First, that it will be an enjoyable read. Second, that it will help increase the reader's practical understanding of using Lean tools by reviewing my actual application experiences. On the enjoyable read side, I have led or participated in the manufacture of everything from 16-foot runabout boats through 164-foot world-class mega-yachts selling for well over $30 million. In addition, I had the pleasure of working at Outboard Marine Corporation (OMC) for three years where they manufactured Johnson and Evinrude boat motors. During my time at OMC, I found myself with an opportunity to travel to Japan and Taiwan to audit the quality processes of Japanese manufacturers, all while I was studying JIT principles. Fun, fun, and more fun!

Here goes…

1

A Manufacturing Family

I grew up in Michigan, about an hour outside of Detroit in a bedroom community. During the 1970s, everyone was working. The automobile industry was hopping, and a lot of people either worked in the car factories (as we called them), or in one of the hundreds, if not thousands, of little automotive component factories supplying the Big Three. My hometown had its share of the little factories on a road leaving the town. I had friends who would say, "My mom makes the little knobs for the handle you roll up car windows with." I had an aunt who made the louvers that went in the rear windows of either Firebirds or Camaros. I had a cousin who worked for a company that made all the tubing that went in cars for brake lines, etc. I had one uncle who worked for Ford, one for Chrysler, and one who worked for General Motors. My grandfather was a crane operator for the Budd Company. He moved pig iron into the foundry furnaces to be made into brake drums that were used by all the car companies. He also died of black lung a year into retirement. Over the years, he would tell me stories about work. The blast furnace, the crane, the hot metal…industry!

Michigan, especially right outside of Detroit, was a fun place to be in the 1970s. In our small town, cars would drive through town, go around a block at each end, and go back down the main street. It was called "taking laps." And the fun thing was the muscle cars on Friday nights. Larry Carpenter's '76 Nova had a 396, hooker headers, Edelbrock high rise, 4-barrel, traction bar, Muncie transmission, Hurst shifter, and a tach held on with a hose clamp. It was a sight to both see and hear go by. Before that, he had a Mustang with many of the same accoutrements. Jeff Denean had a purple '68 Camaro with a white stripe on it, and Randy Hademac had a Corvette with a carburetor about a foot tall sticking out of the hood. Someone had a bright orange Road Runner, and someone had a lime green "Cuda." On Saturday nights, people would go to "Short-cut" road and race. This was a paved road in the middle of nowhere with no houses on it. The guy in the "Cuda" might be racing some kid in his Dad's brand-new F-150 pickup. Man, if his Dad ever knew. You never knew who or what was racing.

So, in the community where I grew up, most people worked at the car companies, in some sort of housing construction trade, and in my father's case, Chris-Craft boats. We were a blue-collar family, and it was great.

While growing up in Michigan, I hunted and fished constantly. I loved it so much that my dream was to be a game warden so I could be in the outdoors all the time. I was going to be in the great outdoors all day, every day. I often told my Dad, "I would never work in a factory." No way. I would not be stuck working inside a big factory. To me, working in a factory was akin to being in prison. How ironic that I have spent most of my life in factories, and enjoyed it immensely.

My Dad was building boats at Chris-Craft boats the day I was born. He had started building boats shortly after high school. He started out as a carpenter in 1955. He loved to work with wood and his hands, was very good at it, and was the take-charge type. He learned most of the jobs. When his crew leader was out, he would take charge and do whatever needed to be done and offer help to others. The other guys would ask him what to do. That led to him becoming a crew leader, and then supervisor. They built the beautiful mahogany runabouts that Chris-Craft was known for.

Chris-Craft was in Algonac, Michigan, a small blue-collar town on the St. Clair River. This was the plant where Christopher Columbus Smith began the legendary boat company. So big in the 1960s that in some countries they used the word Chris-Craft as the word for boat. The plant was located on the river not far from Harsens's Island, named for Chris Smith's son. If you have never seen the St. Clair River on a sunny day in Michigan, you ought to. It is gorgeous. What a river! The water was crystal clear and bright blue, a mile wide, looking over at Ontario, Canada. As a kid, I lived on that river. I had a small boat and fished for walleye. The smell and sound of that river to this day stir powerful emotions in me when I visit there. I learned how to swim in that river, and my lips were blue after those lessons too.

Off of the St. Clair River was a small tributary called the Belle River. It was fed by the drainage ditches and furrows from farmers' fields way on out in the country. Other than in the spring, or after large periods of rain, there was very little current in the river. There were a couple of small marinas, and the rest of the river was lined with private docks. Boat after boat, and beautiful boats too. Boats that we took for granted then. Chris-Craft mahogany runabouts were prevalent, with their beautiful varnished exteriors and red vinyl upholstery. There were all variations of Chris-Craft cabin cruisers, and all made from wood. In one boathouse, there was a fifty-seven footer called the "Sharnell." There was a stream of boats full of families tooling out toward the St. Clair River. Coolers packed with snacks, sandwiches, and what have you.

In the winter, each neighborhood shoveled a pond on the river to play ice hockey on. On Saturdays, we would skate up and down the river looking for a pick-up hockey game. We all had our Eddie Bauer hockey skates (usually a Christmas gift each year) and a hockey stick. Goalies often wore their baseball catcher's garb and padding. The ice would pop loudly as we skated. Sometimes cracks would form and we would catch our skates on them.

In the summer, you never knew what you were going to see coming down that river. Kids, teenagers, and others would assemble old boat motors from anywhere they could get parts or an old motor carcass. It was not unusual to hear the backfire of an out-of-tune boat motor coming down the river in a cloud of smoke that two strokes are known for. Sometimes, the guys in the boat had the cowling off the motor and were turning screws on the carburetors trying to get the motor to smooth out. Boats were everywhere and a big part of life in that community.

In the spring when the snow melted, the river became risky. The water might drop and then rise, depending on the weather. When a rain finally came along with the melt, the ice rose up and away it went out into the St. Clair River, which was already full of ice chunks floating from Lake Huron on into Lake St. Clair. What a beautiful site. Next came smelt fishing. We would go to the St. Clair River with long-handled nets and garbage cans with bags in them. We took turns netting, emptying the smelt into the garbage bags after each dip, or drinking beer and watching. Once the smelt run was over, suckers ran in the Belle River. We fished for them with night crawlers and gave them to an old man who smoked them.

In June, the St. Clair River filled with walleye. I would get up at 3 a.m., and take three dozen or more nightcrawlers that I caught myself with a flashlight at night after a rainstorm and head for my boat. The smell of the river was pungent and awesome. I do not know any other way to describe it. It smells the same today. I would idle down the Belle and out into the "big river." Down at the end of the McGlouth shipping yard, my Dad had told me about an old sugar factory that used to drain into the river. Evidently it formed some sort of structure that attracted walleye. He had pointed out a bush that stood right where the drain used to come out. He told me to line up off of the bush and even with the sea wall about a half-mile up and drop the anchor. I would still-fish there for three or four hours every morning. It was nothing to catch ten nice-size walleye. I would come home at about 9:00 a.m. and clean the fish on our picnic table. I would immediately take the first one in to my mother, who would fry it up for me while I was cleaning the rest. All were filleted, put in milk jugs full of water, and put into the freezer. Moments later, I would be eating the snow-white flaky meat of a walleye that had been swimming around four hours earlier. Delicious!

Another rite of passage in Lower Michigan was spearing spawning fish in drainage ditches. Some people used to say it was illegal. The snow melts and spring rain fills the drainage ditches, helping water make its way to the bigger rivers. Northern pike and carp would swim up into these ditches and spawn. It was not hard to find big pike in the ditches, especially if the water dropped quickly, leaving them in pockets of water. We would ride up and down gravel roads in the country, drinking beer with music playing, alongside the ditches with spears in the car. We would leave the passenger car door open so we could get out quickly. The music would scare the fish and we would see the water ripple. We would stop very quickly and spear the fish. In fairness, our success rate was about one in thirty tries. These fish were wary and not easy to catch.

Last rite of passage. I had a '69 Mustang. It and other cars' wheelbases were the same width as railroad tracks. We would let a little air out of the tires and drive slowly onto the tracks. The tires would lap right over the tracks. We would then put the car in gear, climb out the windows, and sit on top of the car drinking beer while driving down the tracks with nobody behind the wheel…for hours.

OK, one more. In the winter when the Belle River was frozen, if the cops chased us, we would drive onto the river very quickly and turn off the headlights. The police would never drive onto the river. Not always a happy ending; but for me, I was lucky.

Marine City was just a few miles from Algonac, where Chris-Craft was. Dad got up early for work at the Chris-Craft plant for as long as I can remember. He had this lunchbox that could be plugged into an outlet to heat up soup. Mom would grind up bologna and pickles for ground bologna sandwiches. She would bolt the cast-iron grinder to the edge of the table and push in chunks of bologna and sweet pickles. Out would come "ground bologna"! Dad also took a Hostess® Sno Ball® or Suzy Q® cake in his lunch. I would check his lunchbox at night and snatch one if he had not eaten it that day. This was the life of a factory worker, and it was great.

Dad would also bring home mimeographed copies of jokes like the "Polish gas chamber" or something. Some of them crude, but mostly just stupid, but I would give anything to have all of them in a binder right now. It would be a great piece of Americana. Those things would float around the factories and other places for days, and people would laugh and laugh. How simple compared to the twenty-five e-mail jokes we now get each day.

The boat builders at Chris-Craft were talented, very talented. They could do anything with wood. This was evident each year during the Pinewood Derby® at Cub Scouts, a homemade wooden car race. We were all given the proverbial car kit consisting of a small block of wood about ten inches long, four plastic wheels, and four nails for axles. Pretty simple. So there were the boys who carved their own cars. You could sure tell. And that was how it was supposed to be. Then there were the boat builders' kids' cars. Dad would say, "I'll take it to work and 'get it started' for you." A week later, he came home with a sleekly shaped car, beautifully painted shiny blue. Cool! At the race, people would roll their eyes seeing the cars made by the fathers. Oh well, they could have been guilty of worse crimes. Oh yeah, and Dad put graphite on the nails to make it go faster!

Ice fishing was pretty popular in Michigan. The boat builders would carve wooden fish decoys. The decoys were dropped into a hole in the ice under a "fish shanty." They would circle their way to the bottom and rest. The fisherman would pull them back up and drop them again about once every half an hour. The idea was for them to catch the eye of a big northern pike or muskie. When they swam over to investigate, the fisherman would spear them. The decoys were pretty elaborate. The body would be carved out of pine and shaped a bit like a half-moon to get them to circle on the way down. Pieces of tin would be snipped into the shape of fins on the sides and tops of the body and inserted into slits to hold them. A cavity was carved in the bottom and filled with hot lead to give it weight so it would sink. The tin fins could be bent to achieve the best circling action. They were painted all different colors, usually to mimic a yellow perch or some sort of shad.

Chris-Craft was clearly ahead of its time in manufacturing. The plants were clean and highly organized, with a place for everything and everything in its place. Very much like the result of the Japanese 5S (sorting, straightening, systematic cleaning, standardizing, and sustaining) philosophy of managing a work area. Pride was evident everywhere. The builders had handmade toolboxes with jack planes, box planes, braces, bits, levels, and calipers. Maybe even a "Vargas girl" picture varnished on the end.

Each builder knew the purpose of each tool and how to use it. Garage, yard, and estate sales in the area always had a complement of these tools. To this day, you can find those tools if you care to take the time to look.

Even in the 1960s before materials requirements planning (MRP) and enterprise resource planning (ERP) and all the other manufacturing programs we have today, Chris-Craft had a part numbering system still referred to today for replacement parts. They even stamped part numbers on the back of different wood parts on a boat! They did a better job of managing plant inventories than some of the most well-equipped plants thirty years later.

In 1969, Chris-Craft closed the Algonac plant. This was quite a blow to the boat builders and the local community, but it was still during the time of a manufacturing economy so the boat builders all found jobs in other trades. Luckily, there was a relative abundance of automotive and other jobs in Michigan at the time. Still, this would be a significant change in the lives of the boat builders and their families, and one of them was in mine. Boat building is a very unique profession. There was a lot of hand labor and close interaction with people. The components made of wood or fiberglass for boats did not have the tolerances of the machined parts used in cars or other types of manufacturing. In a typical boat plant, there are mold makers, fiberglass repairmen, painters, sprayers, engine installers, woodworkers, cabinet builders, electricians, upholsterers, assemblers, welders, and canvas makers, all detailed handwork requiring skill and attention to detail.

I was in fifth grade when the plant closed. It was my first plant closing. Little did I know that I would live to see many, many other boat plants close later in my life, and the impact they would have on my life and that of others. On the other hand, I saw many other plants open. I have stood in the middle of thriving boat plants with hundreds of boat builders just building away. The sounds of air guns, drills, saws, hoists, forklifts, and things being pushed along were captivating. And then, I have stood in an empty plant too, listening to the eerie silence. In the early 1990s, I made a trip to Michigan to see an old friend. I made it over to Algonac where the original Chris-Craft plant was. It had been turned into a marina, and the building was being used for boat storage. I stopped by on a whim and asked the marina manager if I could go inside the building. I don't know, call me sentimental or whatever; but it was really moving to me to stand in a factory that had produced thousands of mahogany runabouts and Catalinas® and other fine boats. I could almost smell the sawdust and hear the tools as the boat builders built twenty-five years earlier. The building is there to this day. I know that this is life, but I get sad when I see an abandoned factory, especially when I think of the little community of people who worked in it.

So away life went and one day Dad told us that he got a job with a company called Sea Ray®. Sea Ray was located Oxford, Michigan, about forty-eight miles from where we lived. Dad drove the ninety-six-mile round-trip everyday to work. He put a lot of miles on our Chevy half-ton pickup. I continued on with growing up…school, sports, hunting. Once a year, we all went to the Sea Ray company picnic. The company put six to eight new boats in the water, and all the supervisors took turns giving rides. As many rides as you wanted. Hot dogs, hamburgers, and other food was everywhere. It was a blast! One year they even put fish in the factory test tank for a fishing pond! My brother and I always enjoyed the ride on the Sea Ray Pachanga®, their version of a speedboat. It had a 454 Berkeley jet drive that made the boat take off like a rocket. The torque at take-off pushed your head against the seat, and did it big time. What fun!

Several years later, Mom and Dad decided to build a home near the plant, which meant we would be moving. We moved in with my grandparents on their farm until the house was complete. The timing was terrible because it resulted in a challenge to my life. This challenge was that I had to move away for my senior year of high school. My dad gave me the option of staying with my grandparents, but giving my brother and me a new home with our own rooms was something he wanted to do for us. The timing was bad, that was all. So away we went.

It was a tough year, yet a lot of fun. Bittersweet. I missed my girlfriend and friends, and the high school in Lapeer was much bigger. It was tough making new friends. But I did, and managed to make a few. I got in trouble at school, and my dad got called at the boat plant to come down to the school. I could just picture my dad on a boat working with boat builders in getting a fuel tank to fit, or cutting down a bulkhead to get a galley in. The vice principal told me I needed to call my parents. I told them that Dad worked at Sea Ray boats and they called. He was out on the production floor when he got the page: "Mr. Trudell, you need to come down to the school. Your son is in trouble." Wow, as the father of two sons, I now know the joys of raising sons. They test your patience. Many times as a plant manager myself, I smile and think of Dad getting that call when I was walking around the factory.

Lessons Learned

- Factory workers and employees have a tremendous amount of skill, pride, and talent.
- Factory workers are people, and things like plant closings impact not just them, but also their families and their lives in big ways.
- If you want something, be willing to work hard for it.

2

My First Factory

I graduated high school in June 1975 and immediately got a job at a gas station. It was my first job out of school. I had worked part-time at restaurants washing dishes, etc., but I was not out of high school and this was my start. It was a fun job. "Fill 'er up," "Check the oil," "Give me $20 of regular." Oh yeah, and every car with a pretty girl got the windshield cleaned! So, it was going good, for a week. Then one day the owner tells me that the guy who I had replaced wanted his old job back and he felt obligated to give it to him because he had worked there for several years. So, I was unemployed.

It was 5:00 a.m. about a week later when the phone at my friend's house rang. We had only gone to sleep a couple of hours earlier after a night of partying. My friend nudged me and said, "It's for you. It's your dad." My dad? Hello? "If you want a job, get in here now." He was calling from Sea Ray®. He was the General Foreman of most of the factory.

I took a quick shower, grabbed a coffee, and headed out the door. I was tired and hung over. It was a long forty-eight miles and I barely knew the way. There was no direct route. Marine City Highway to Palms Road to 28 Mile Road to Richmond. Left in Richmond on Romeo Road and all the way down Romeo to Lakeville. Lakeville Road to Oxford, and Oxford Road to the plant.

I showed up at the plant and went to HR as Dad had told me to do. A man named Ike Morrisey was the HR manager. He advised me that I would be in the gelcoating department, which consisted of two men, Randy and Dave, who gelcoated all the molds the boats were made in. Gelcoat is the "paint" on fiberglass boats. The difference is that the first thing you do to a boat is put the outside finish on it—as opposed to a car, where it comes after the steel is made. The gelcoat needs to be correct. It is very costly and disruptive to the plant if it is not. Because of this, the gelcoaters were the highest-paid boat builders in the plant. They were "the men" too. I was to be their "attendant." Mr. Morrisey added that he thought they had a "project" for me to do before starting in the spray booth.

HR asked me to fill out some paperwork. I finished about break time so they took me to the break area and told me to come back after break and they would introduce me to the gelcoaters. The break area was full of boat builders. I was eighteen years old. They were grown men. It was an awakening to hear them cussing and swearing and talking shop while smoking and playing cards. There was a lot of smack talking. The atmosphere was surreal to a young guy. You could smell food, but no microwaves, as it was 1975. All kinds of lunchboxes with sandwiches and leftover foods in Tupperware®-type containers. And everyone had a Stanley® thermos, the light green kind. Most with some sort of handle affixed to them with a big hose clamp. I still have mine today. I was still hung over and tired. I decided to lay on top of a picnic table to rest my eyes.

I kept feeling something tugging on me and a voice saying my name. Holy smoke! I fell asleep on the table! My new supervisor was waking me up. "Hey, are you planning on working?" he asked me. I explained that I got called on short notice. He said, "Well if you want to work here, I suggest you be where you're supposed to be." What a great start—not.

The hull and deck molds are shaped like the parts made in them, only a little bigger. They had piping laminated on the outside to give them strength and ten-inch steel wheels on the bottom so they could be moved. The company hired big football player sized guys to push the molds from station to station in the factory. Boats were built progressively from station to station where different layers of fiberglass and miscellaneous wood and stiffeners are laminated into them. They get thicker and thicker through the process.

Basically two different types of fiberglass were put into the boats. After the gelcoat was sprayed in the mold, fiberglass "chop" was applied. An air-powered chopper gun had a rubber wheel that pulled fiberglass twine from spools in boxes. There were razor-type blades that spun and cut the fiberglass into strands about 1½ inches long and spit it forward. The strands would be covered with fiberglass resin sprayed from a tip at the bottom of the gun. The glass strands and the resin would meet in the air and fall onto the part.

"Laminators" would roll out the chop with rollers that looked like a spiny paint roller. The idea was to roll the air out of the chop and tuck it in all the nooks and crannies in the molds. This was detailed work because any place there was air would result in auto body type repairs having to be made after the boat was out of the mold. It just so happened that there were a lot of women in the lamination department. Some speculated that women were better at detailed and tedious work. For whatever reason, maybe just happenstance, many of these women were rather "large." The supervisor of that department was named Omer, but people called him Homer. Thus, the "large" ladies were called "Homer's heavies." You have to love factories.

OK, so the big football player guys, called "mold movers," pushed the hull, deck, and liner molds from station to station where the fiberglass chop was applied. The excess chop would build up on the outside of the molds after a few months, and it would often reach about a foot thick. The wheels would clog up with chop and fiberglass residue and become hard for the mold movers to push.

This is where my new job came in. My first project was to scrape the chop off the molds. I think they gave that job to new people. I would also jack them up and remove the wheels to grind the residue off of them and grease the bearings. What a job! I used a chisel welded to a piece of pipe with a T-handle on the end to do the scraping. I got fiberglass all over myself while working. I never knew what itching was until I had this job. I quickly learned to wear long sleeves and put tape around the cuffs.

I finished all the molds in about a month. I then started at the gelcoat spray booth as the spray booth attendant. My job was to wax the brass rails on the edge of the mold and mix the gelcoat for the gelcoaters. However, there was a great deal of residual gelcoat on the rails that had to be scraped off with a chisel before the wax could be applied. Next, we would look on the order to see what color the boat was, and I would put that color in the spray pot with the proper amount of catalyst. The catalyst was methylethylketoneperoxide dimethylphthalate. It was the craziest name of anything I had ever seen, so I memorized it. After adding the catalyst, I would mix it with an air drill with a mixing tip on the end. I would give the sprayers the sign and they would spray. Away they would go, moving the spray gun back and forth to evenly apply the gel. The sprayers wore paper suits, with tape around their shoes, and gloves, a hood, and a filtered mask.

The gelcoat would begin to catalyze in about ten to fifteen minutes; so as soon as they were done, I would have to clean the spray pot. Timing was key, because if you did not pay attention, the gelcoat would catalyze in the hoses and that meant throwing them away and a lot of time and trouble to set up new ones. Now this was manufacturing in the 1970s. First we kept one or two 55-gallon drums of acetone in the area. I would fill the gelcoat pot with acetone and then hook up the air and flush the lines of the spray gun. Then I would clean the gun. The gelcoat-stained acetone would be sprayed into a container with a lid on it. That was it. We did about eight to ten boats per day—so eight to ten decks, eight to ten hulls, eight to ten liners, and eight to ten sets of small parts each and every day.

Once when I was cleaning up, I thought that inserting the tip of the mixer blades into the used acetone and spinning it would be a good way to remove the gelcoat from the blades. Little did I know there were some cleaning rags in the bottom of the container. When I spun the mixer, the rags grabbed a hold and dirty gelcoat covered my face and eyes. It hurt. I washed my eyes with water. I then wiped down with acetone. That was another thing. Each day at the end of shift, we would take some clean cotton rags and wet them with acetone. Then we would wipe down our arms, face, and hands. We would stand in front of a fan while doing this, and the acetone would feel like air conditioning as it evaporated. Thirty years later, acetone is not even allowed in most factories, let alone in fifty-five-gallon drums, or even allowed to be used to wipe down our bodies. Things have changed a lot.

After a few months, I was advised that there was an opening in the mold maintenance department for a mold repair specialist. Mold repair specialists were in one of the higher-paying categories at the plant, so I applied. I got the transfer and was immediately assigned to waxing molds. I later began to learn mold repair. My new group leader's name was Bill Bush. He stood about six-foot, six-inches tall, was skinny, and had red hair with a comb-over. Man, was he a worker! He could patch, re-spray, buff, and his work was incredible; and he expected that from us too. He worked like an athlete and he was quite a sight to behold. His comb-over would go over to one side and his tongue would hang out. We would laugh at him, but we also greatly respected him. We would work for hours repairing molds only for him to mark them up all over with a yellow "China marker." There was more to do when he got done than when we started. He was tough. On the other hand, he was a real team leader. He was the best repairman we had, and he led by example. When he was finished with a mold, it was beautiful.

In the 1970s, the polyester molds that boats were laminated in were not made of the better resins and materials in use today. Combine that with the lower-performing releasing agents and gelcoats, and the molds took a beating. You could often hear the thumps of rubber mallets that the parts pullers used to get parts out of the mold. They worked like chiropractors to get out the parts. In the process, the molds took a beating. The molds often cracked from the force of pounding of the mallets during pulling. Cracks occurred from dropping the mold against the floor to break the parts free and from the twisting and further jarring. The gelcoat would often stick to the molds. Chunks of the mold would come off and onto the part coming out. Molds were highly susceptible to human sweat. There were often areas of white gelcoat stuck to the mold in the shapes of handprints or fingerprints. If anyone touched a mold after waxing, there was a good chance the part would stick.

Reworking hull molds was a big job. We would mark them up, do all the patches, then the re-sprays, then sand and buff. We held eighteen-pound buffers over the sides of hulls buffing away for hours. We had these orange bricks of buffing compound that we would run our buffing pads against before buffing the part. You had to pay attention, however, because if you buffed too long on one spot, you would "burn through" the mold. If that happened, you got a good "chewing-out" from Mr. Bush, and he could give a good one. If he were working today, he would spend a lot of time in HR, and I am not sure that is a good thing. But we learned a lot from him and he expected a lot from us.

The boat factory was pretty big. It was nothing like an automotive plant, but about 300,000 square feet and about 650 employees. It was a really fun atmosphere. There were lots of young people, and some of them were girls, which was great for me as a young man. There was Cheryl, a cute redhead on the runabout line. There was Lolita, the short and cute girl in final finish. Then there were the sisters on the cruiser line—Anita, Cindi, and Mary—always flirting. The factory was exciting: forklifts transporting materials everywhere, all sorts of handtools being used—drills, screw guns, impact wrenches, hoists, glue guns, saws, routers, hammers, and you name it. It was like a small city. And unlike anything I had ever seen.

The experience of living the life of a factory worker was great. I made a lot of great friends, had fun while working, and hey, I got paid every week. While driving to work each day, I passed a Pontiac dealership in a small town. They kept it lit up all night for security reasons. One morning, I noticed a brand-new silver 1975 Firebird in the showroom all lit up. I stopped and peered at it through the glass. It was awesome. I was enamored. I was thinking like, "Oh my gosh, wouldn't it be cool to have a car like that?"

A few weeks later, I told my mother about the car. To my surprise, she said, "Hey, you're working. If you want to buy a new car, you can. You just have to make the payments and pay the insurance and gas." That was all it took. Two weeks later, my parents took me out to dinner on the way to the dealership to pick it up, and I was so excited I could not eat. What a great way to spend $4,850. It was silver with a blue racing stripe, wood-grain dash, and rally wheels. I took it the next weekend and had a Craig 8-track power play installed in it. Too cool!

Then there were the day-to-day charades of factory workers. This caught my attention immediately. Being somewhat of a jokester myself, I enjoyed this immensely. Take, for instance, a prank often pulled on mold waxers. Before waxing a mold, we would take off our work boots and wrap our feet in clean white cotton rags to keep the sweat from our feet from coming into contact with the molds. Then we would crawl inside the hull molds and wax them, much like waxing a car. Wipe on, wipe off. Polyester putty is abundantly used in a boat plant and we would often, upon exiting the mold, find our boots puttied to the floor! Lots of cat-calls and hoots would ensue.

On occasion I chuckle when reflecting upon another instance of pranksters in the plant. The mold maintenance department was a four-walled room creating an area to hold all the molds and do necessary maintenance and repair. The walls were concrete block, and at one end of the department there was about a two-inch hole that had been drilled or wallowed out. You could peek through and see production workers in another department. Usually, this hole was full of dust, and also contained a cavalcade of Popsicle® sticks used for patch paste, screws, fasteners, and other miscellaneous debris that factory workers would stuff into the hole for unknown reasons. One day, one of the mold repair workers inadvertently peeked through the hole and proclaimed that the vice president of operations and the plant manager were on the other side talking with workers. So what did he do? He proceeded to stick an air nozzle in the hole and let the air fly. On the other side, a tornado of debris and dust went flying out of the hole and engulfed those VIPs. Holy smoke! Everyone grabbed some sandpaper or a buffer and worked feverishly. A few minutes later, the executives and our supervisor entered our department. We were the busiest group of workers you have ever seen! Not so much as a smirk occurred.

As I learned mold repair, the pressure on the department began to increase daily. I would get up at 3:00 a.m., drive forty-eight miles to work, clock in at 4:00 a.m., and work until 6:30 p.m. at night, plus a lot of Saturdays. On the bright side, I was making a lot of money. Hey, after all, I earned somewhere around $4.50 an hour, and that meant $6.75 per hour for overtime. That was big bucks! On the downside, I was tired all the time from working all those long hours and participating in all the partying that goes on when you are eighteen years old. I would fall asleep at the drive-in movies during dates, and that did not go over too well. The hours were wearing on me, and I was getting disillusioned with "factory life." I mean, I really began to enjoy working at this factory, but seventy-five hours a week was becoming a bit much. Then the week before Christmas, our supervisor called a meeting and advised us that while the rest of the plant was closing for two weeks over the holidays, our department was working mandatory twelve hours a day for the whole shutdown. We would be off on Christmas Eve and Christmas Day, and New Years Eve and Day, and twelve hours of work every other day. This was too much for me. I headed to HR and advised them that I liked working there but that I could no longer work those types of hours. I was willing to work fifty or sixty hours a week, but not relentless eighty-plus hours a week. I wanted a life. HR told me to go home and think about it.

My phone rang during the first week of January. It was HR. They advised me to come in as they had a job on the first station cruiser line as a liner installer. More specifically, this job entailed prepping a fiberglass floor liner, applying "fur" to the bottom of it, which was a ceiling in a mid-cabin after installation. Once the engine was installed in the 240 Sundancer® cruisers, I lifted the liner into the air with a strap and a hoist. It was a wobbly and tricky lift. One false move and the liner fell with a big bang, and being in the middle of the department where all could see any misstep made it a little nerve-wracking.

Once the fit was verified, I mixed a bucket of putty with the catalyst. I then shimmed up the liner with masonite shims and puttied it into the boat. Not a bad job. To get the putty, I walked to another department where there was a fifty-five-gallon drum of putty. I would reach down into the barrel and scoop up putty in the bucket. If you were careful, you could keep from getting putty on yourself most of the time. When I mixed the catalyst into the putty, I would sit on a wooden box with carpet on it, hold the paper gallon bucket between my ankles, and stir with a wooden stick. I would use it to set the liner. One day, as a prank, someone poured catalyst on the carpet on my box stool. After mixing a batch of putty, I felt a burning sensation on my behind. I felt the chemical on my jeans and I smelled it. It was definitely catalyst. This was very much akin to sitting on a cushion soaked with battery acid. I had to drive home and wash my behind. I was left with a "nice big cherry" on my backside. Oh what fun! Ha, ha. I was back at work two hours later.

Back then, the job that caught my eye was that of "engine installer." There were about seven or eight assembly lines in the plant, and each line had and engine installer. The engine installer was the "guy job" of jobs. "Tim the Tool Man" would definitely not want this job. This job entailed prepping the engines by putting oil in them, hooking up the battery cables, installing associated harnesses and cables, and installing the engines in the boat. Wow! Brand-new engines, lots of wrenches and tools. Hubba, hubba!

This job came open and I immediately applied. I got the job. Oh man! The previous engine installer, Alan Peters, had been selected to be a "roustabout," which was a job that included filling in for whomever was absent that day. Obviously it took a lot of job knowledge, and this job often led to a promotion to group leader. The good news was that Alan would train me for two weeks before moving into his new position. Five boats a day with two engines each for two weeks meant a hundred engine installations of training.

I was now on the cusp of becoming' one of "the few and the proud"'— an engine installer! Everyone knew all the installers. Joe Hughes was the best. Joe could install an engine after set-up in eighteen minutes. This was a big deal. His sister Teresa also worked in the department and she was quite a "looker." On occasion when the timing of the lines was right where more than one engine was to be installed about the same time, the engine installers would compete to see who could install the engine the fastest. After the boat and engines were prepped, the installers would hoist the engines in the air just over the boat. They would wait until the second hand on the department clock reached "12" and off they went. The whole department would cheer them on. It was great fun and rivalry. Imagine that! Workers competing to see who was the fastest. Believe it or not, quality—and high quality—was expected too. Back then, people took pride in being a good worker. It was a real badge of honor.

At eighteen years old, this was where I experienced my first lesson about quality. Alan Peters worked like a robot. He was a respected worker known for speed, quality, and having the knowledge to do any job. Man, was he a good worker! He advised me to just watch him work for a couple of days. He explained the job to me all the while he worked. He worked like a member of a pit crew. He explained to me that we have to "set up." "If you do it right, the job goes a lot faster." First he said, "Go to the stockroom and get ten battery cables each day at the end of shift: five red, five black. Then you will be prepped for the next day." This was standard work! In

1975! Every day, every engine was installed in exactly the same way. The same order, the same drill, the same impact wrench, the same way. The only changes were for improvement.

There were ten brand-new engines in my area upon arrival the next morning. Alan immediately installed the battery cables to each one. He then noted if the boat was a "dual" station and if it was, we installed a dual-station sending unit so the oil pressure could be read at both helms on the boat. We then proceeded to fill the engines with oil from a fifty-five-gallon drum. It had a dial on it and we set it at 5 or 6 quarts and pulled the trigger until it stopped clicking. We then took the nuts off all the bolts holding the engines to the oak skids they came on.

Next, Alan said, "Watch this." He grabbed three yellow barrels, six blue barrels, and two eyes, and set them inside the boat on the floor shelf. Following that, he reached over and put a specific wrench on one side and crimpers, etc., on the other side. He put each tool in a place right next to where it would be needed during the installation process. Once the boat was set up, he hoisted an engine into the air. He then jumped in the boat and, working as quickly as he could, he installed everything. He then picked up the trash and paged the QA inspector to "Engine check, first station cruiser line!" A few minutes later, a QA Inspector appeared and verified that the engine was properly aligned and a couple of other critical tasks had been performed properly.

Here was my standard work (none of it was written down, by the way):

- Verify that all engines in area are the correct ones for the day by reviewing the sales orders on the boats coming up the line. (The day's boats were usually in a row and you could see them. There were other operations being done to them, such as having the holes cut in them, the fuel tanks installed, etc.)
- Put oil in all engines. There was a fifty-five-gallon drum of oil with an air pump on it.
- Install battery cables—one red and one black—on each engine.
- Install extra oil, water, and temp sending units if it is a bridge boat.
- Remove the nuts from the bolts holding the engines to shipping crates.
- Go to the stockroom and get ten shifter cables, ten throttle cables, five steering cables, forty lag bolts, and some masonite shims.
- Go get a one-gallon bucket of putty for the day.
- Go get ten foam blocks (like giant loaves of bread) from the foaming department.

Each time a boat moved into my station:

- Put barrel connectors on shelves inside the boat, along with four lag bolts on each side.
- Throw the cables in the boat, along with some shims, metal strapping.
- Hoist up an engine.
- Get in the boat and lower it to the gimble.
- Install the gimble bolts.
- Get the other engine and do the same thing.
- Get out of the boat and insert the engine alignment spline inside the outdrive port. Lift the engine up and down until it slides in and out easily.
- Adjust the engine mounts down a certain amount and fill rest of gap between mount and stringers with masonite shims.
- Lower the engine's weight on the shims.
- Re-check alignment.
- Drill holes for engine lags through the mounts.
- Mix small amount of putty and put in between shims and stringer.
- Use the impact wrench to tighten down the lags.
- Get out of the boat and re-check the shaft alignment. If good, go to page phone and announce over the PA, "Engine check, first station cruiser line."
- Get back into the boat and install all cables and the engine harness.
- Quality auditor verifies alignment and initials the sales order.
- Strap in the foam blocks on either side of the engines.
- Pick up your tools, get out of the boat, and get ready for the next boat.

Five times per day, I accomplished this routine, with only slight variation when engine brands changed from Mercruiser®, to OMC® or Volvo Penta®.

I had now made it through a couple of boats. I made it through installing a set of twin engines. Man, great! I was proud. I was ready to page QA when Alan suggested that he review my work. Upon looking it over, he pointed out that the wire ties I installed were randomly placed, as opposed to equidistant apart. Although the engine was aligned with the gimble, there was some friction with the alignment tool, but not much. The paint I applied to the floatation blocks was spotty in a couple of places. All in all, however, it was not bad. I excitedly confirmed to Alan that he could count on the next boat being completely right. It was then that he advised me that, no, *this* one would be completely right. He made me undo all that was wrong and do it over. "It is either right or it is wrong, Bill," he said.

"There is no in-between, and we get paid to do it right." I redid my work, and I cannot tell you how many times I have recalled that moment when talking to other operators as a plant manager. I have appreciated Alan for holding me accountable for quality ever since; and if I ever were to see him again, I will thank him.

In 1975 at Sea Ray Boats in Oxford, Michigan, quality was a huge deal. It was absolutely expected. All boats were tested at the "test tank" and inspected for quality. The worst thing that could happen to you is to hear your name over the PA system: "Bill Trudell, please report to the test tank." The one time it did, I heard cat-calls: "Hey, Willy! You need me to do it for you?" That was not an experience you wanted to have. So again, it was a badge of honor to be a fast worker, and quality was assumed.

The other behavior that was in place was that the first thing that everyone did upon receiving any component or sub-assembly was to check it for quality. If anything was wrong, they immediately contacted the person creating the defect and that person immediately addressed it. Again, there was no "I'll get it right on the next one."

During this time as an engine installer, I really began to enjoy the factory and all the fun stuff that went along with it. There were colorful people, and a lot of them. The factory had somewhere around 600 employees and, oh yes, some of them were girls. Cheryl Brown, the cute redhead on the runabout line. Lolita Hart, the tiny brunette who worked in final finish. And then there were the Simmers sisters. Anita, blonde and beautiful; Cindy, brunette and beautiful; and there were more...Kay, Arlene, ... These girls wore tight jeans and were a blast to work with and pretty to observe, especially as a young man. One day they bought me cologne. So this was work! On Saturday night, the local bar, the Royal Lounge, was full of workers from the boat factory and it was just a blast.

It was during this time that I really became enamored with factories. I was constantly looking around at all the people and the fun we had working together, and it was just great. The pranks, camaraderie, and friendships were awesome. For instance, there was a rather well-blessed girl with red hair known as "Big Red"; her real name was Doris. She was a very tall and attractive girl who installed the headliner on the underside of boat decks. About once a month she would wear a bright yellow shirt that said "boobs!" on it. Man, the whole factory would be in an uproar. Constant energy and buzz all day long! Well, finally one day when she wore it, there was a commotion in the department next door. A guy named Johnny Malloy retrieved something from his toolbox and quickly adorned a

T-shirt that said "balls!" on it. He promptly went over by Doris and put his arm around her. Everyone had a big laugh, and it was all in fun. I think someone took a picture. If that happened today, HR would no doubt be in an uproar.

As time went on, I noticed the job of "plant manager." Our plant manager, Doug Slater, was a pretty good guy. He came into the factory every now and then, and interacted with the employees. He would page people regularly during the day, "So and so, please call extension 31," which was his telephone extension number. What I saw was that the plant manager had control over his schedule, got to lead people, could change things in the factory at will, had a company boat, and made a good salary. I mean, the use of a brand-new boat each year for the summer. That was awesome. It looked like a great job to me, and I was thinking more and more that this was the job that I wanted. It was clear that to get this job, I would have to go to college, which meant I had to stop partying and get some self-discipline. So, one day I decided to join the U.S. Army so I could get away from my partying friends and find a way to go to college. So I went to Hawaii for three years and, in return, they paid most of my way through college. I got a reserve commission while in ROTC to earn some extra money while going to college at the University of Tennessee. After being enlisted, it felt so good to be a lieutenant that I decided to go back in as an officer. I did, and earned my MBA going to night school, for which they paid the majority of the costs. What a deal!

Lessons Learned

- Get it right the first time.
- If it isn't right, it isn't done.

3

Maybe Not Lean, But Lean Things

During my second tour in the U.S. Army was where I really gained an understanding of the passion thing. The urge to be a plant manager never left me. My fellow officers were so passionate about being soldiers. They would stay up at night and study FM 7-6 infantry operations or MOUT (military operations in urban terrain). I stayed up late and studied business and manufacturing. I loved the Army, especially the life of being a soldier. The opportunities for personal growth, leadership, discipline, and camaraderie were very meaningful. But I was not passionate about it. It was just fun for me, but that was not enough. What I was passionate about was manufacturing. I could not read enough. I had actually taken several sets of engine installation documents that were usually thrown away from three different marine engine brands and studied them over and over while in the Army.

At the end of four years, I advised my Dad that I was leaving the Army. I remember it was during lunch on the pier in Flagler Beach, Florida. I said, "Well, I have made the decision to get out of the Army." He asked, "So what is it you think you are going to do when you get out?" I said, "I am going to be the plant manager of a boat factory." "And just how do you think you are going to do that?" he asked. I replied, "I don't know. I just am." He was a little miffed, as he had always wanted a military career, and I think he was living out some of that in me. A few weeks later, an envelope from my Dad arrived with some contacts in the boat business.

Several months later, I finished my MBA. Eight weeks before getting out of the Army, I sent a letter to the vice president of manufacturing for Sea Ray® and advised him that I was graduating and getting out of the Army and was hoping for an opportunity. A week later, I received a letter asking me to call him at my convenience. I did and arranged an interview.

I showed up in a navy-blue suit, white shirt, Cross® pen, black leather briefcase, and wing-tip shoes! He offered me an opportunity to participate in a management-training program. He asked me where I wanted to work and how much I wanted to make. I told him. He assigned me at the factory I wanted and set my salary at $3,000 more than what I had asked for. He then said, "OK, I put you where you wanted to be and paid you more than you asked for. I shouldn't hear any complaints from you, should I?"

I spent several weeks in each production department, studying the operation and learning all I could. Before I could progress to another department, I had to give him a tour and explain in detail all the processes in the department and what was key and what was not. What a blast! This just continued to build on my actual experience as a boat builder. Near the end of the program, the VP advised me that I would be assigned as a warehouse supervisor at one of the plants up the road. He told me that it was "a pretty messed up operation" and they needed to get ready to go on MRP. He then said that, "If you clean up this operation and make it a success, we'll promote you to manager, and then you can get some experience and work toward your goal of being a plant manager.

DO IT RIGHT THE FIRST TIME

My background had already included a few years of hands-on boat assembly and materials management, and it was at Sea Ray. Sea Ray, founded by Cornelius Ray (Connie), I think did a lot of things right by accident. Maybe it was just common sense, and I mean no disrespect, but they seemed to just do things right naturally. A lot of their practices were right in alignment with Lean principles. I consider myself lucky to have started out my adult working career there. There was no other way to do anything at Sea Ray than doing it right. If you did it wrong, you did it over; and if you continually did it wrong, you did not work there anymore. The other thing, and this is just my opinion, is that as Sea Ray grew, they only did things if they had the time to do it right. If they started a new plant, they started training a new plant manager in a high-performing existing plant and sometimes some process managers, too. This was before breaking ground for the new facility. By the time the new facility was done, they had a well-trained core staff to start out with.

When they started the plant, they only went at the rate at which they could produce quality boats. If that meant only having one assembly line and running the factory under break-even, then so be it. But the one thing that did not happen was producing a poor-quality boat. Quality was a huge part of the culture at Sea Ray. Quality was expected in every way, shape, and form—period. They drilled it into me and it has stuck with me my whole career. Sea Ray was a great company and is a good company today. To me, the main difference is that back then it was a more personal company, and now they have all the challenges that go along with being part of a big corporation. This comment is not meant to be a cheap shot. I think it is just a reality of today's business environment. There are some people at Sea Ray as I write this who I have a lot of respect for and they still make a great boat.

PRODUCT DEVELOPMENT

The processes for product development, product launch, and new model introduction are the subjects of many books. In any manufacturing environment, producing product to Takt times, standard work, line balance, and level-loaded work centers, a flawed new model introduction can put the plant on its knees. It is basically akin to pouring water in your car's carburetor while it is running. The results will be very similar to what happens to the factory.

I credit the product development and new model introduction processes that Sea Ray has followed over the years as a significant contribution to that company's success. It is partly due to the "they didn't do things any faster than they could do them right" value in their culture. Poor new product integration has cost the U.S. marine industry millions of dollars in wasted labor hours, warranty costs, and lost customers. It has also cost some people their jobs.

Over the years, Sea Ray followed a basic process with some variation from time to time for introducing new product. Once design and product development scoped a new product and created tooling, PD & E created documentation and bills of material for the new model. Every couple of weeks, the receiving plant would send their manufacturing engineers (MEs), who were highly skilled, hands-on boat builders (usually not degreed), to assist PD & E in building the prototype. These MEs had high levels of input into how the boat was to be assembled, and integrated their ideas and preferences into the prototype. As the prototype was completed, the tooling and documentation would be shipped to the manufacturing plant where they would begin to build the second prototype or pre-production unit. I have heard that on some occasions the plant finished the pre-production unit before or close to PD & E finishing the first one. Once the pre-production unit was complete, the laminated parts were sent through the normal laminating processes and an assembly line was set up in the assembly building. The line was staffed with boat builders, and the manufacturing engineering team worked on the line with the boat builders to take the line up to the Takt time balanced with customer demand. If necessary, PD & E staff would also participate in the plant if needed during the introduction. This one paragraph of information, combined with product designed for manufacture and assembly, is one of Sea Ray's cornerstones of success. I describe later what happens when this does not happen.

In Lean, learning and teaching are important values. In my experience, Sea Ray did well in this area. The way to succeed there was to learn as many jobs as you could, be good at all of them, and be willing to do any of them at any time. If a good boat builder learned many jobs, he or she moved into a "roustabout" position where they filled in for absent employees. They, in effect, could do any job in any area. Roustabouts were key candidates to become a team leader or supervisor. With their hands-on knowledge of all the jobs, they could train, teach, and lead operators effectively.

All Sea Ray plants had a lamination department where the boat's superstructure was created out of fiberglass, resin, gelcoat, coring materials, and adhesives. They had a fabrication (sub-assembly) department that made dash pods, seats, cushions, upholstery, radar wings, cabinetry, panels, harnesses, and kits. They had a final assembly department set up with assembly lines, each with its own individual Takt time. They had a stockroom or warehouse. They worked really hard to develop their department or process managers. After one to two years, a manager would swap out with another manager—almost like musical chairs. So for a mature plant, the assembly process manager most likely had a couple of years under his or her belt as a lamination process manager, fabrication process manager, and manufacturing engineering manager, and in many cases had many years as a boat builder. That also made for a pretty strong leadership team and created a solid pool of plant manager candidates. This is right in line with the Lean value of learning many jobs.

PROBLEM SOLVING

Another part of the Sea Ray culture was aggressive problem solving. This is right in line with Lean thinking. Sea Ray did not use a Toyota A3, but did place big value on solving problems quickly in order to prevent delay and delivering product to their customers on time. I will note that Sea Ray had several plants and there was some variability between how the plants operated but in my opinion, there were more similarities than differences. Most all of them knew "the way." I spent the summer at one of the plants in the management training program, and one of the things I observed was the emphasis on quick problem solving. All managers had a Motorola radio on their hip. If a call went out signaling a problem in the process, then within a matter of moments the plant manager, manufacturing engineering manager, department manager, and supervisor of the area with the problem were standing on the plant floor next to the problem. Given the way Sea Ray trained their managers, there was a good chance that each one standing there had managed that area at one time and they had valuable knowledge to use toward solving that problem. They stayed there and dug in as a team, and did not leave until the problem was solved. If it was not solvable, they stopped the line or "dropped a boat" from the schedule to compensate for the lost time.

Note: Be careful when wearing a two-way radio. On one occasion, a supervisor was leaning up against a boat and unknowingly keying his radio's microphone. He had quite a bit on his mind about his boss and other managers, and it was all being broadcast across about twenty-five radios in the plant. Yikes!

I actually had roots in the Knoxville (Tennessee) factory. My father had led its start-up in 1978. He was the general foreman at Sea Ray's Oxford, Michigan, plant in the late 1970s and when Sea Ray decided to start a plant in Knoxville, they asked him to be the plant manager. He set up the plant and in 1979, between a combination of company politics, an idiot, and a bad economy, he left that plant. Many years later, he retired as the engineering quality manager from Sea Ray's Palm Coast, Florida, facility, one of the world's largest and most productive boat manufacturing plants. In the early 2000s, Sea Ray boat sales were over $1 billion. It was the most formidable boat manufacturer in the world.

TAKT TIME

One of the things that amazed me the most was the discipline that Sea Ray applied toward managing schedules. Everything—and I mean everything—was scheduled in the plant. No one was allowed to get behind schedule, and no one was allowed to get ahead of schedule. If they did get behind and could not recover, the schedule was adjusted and moved back to the Takt time. If necessary, the line would work on Saturday to make up the unit. There were boat schedules, helm seat schedules, computer numerically controlled (CNC) schedules, upholstery schedules, helm seat schedules, harness schedules, test schedules, kit cart schedules, small parts schedules...everything was scheduled.

If you went to the CNC area where the wooden stringers were cut, you could walk over to lamination and that would be the boat that was in the gelcoat booth being gelcoated. In two more Takt time cycles, the assembled stringers would be delivered at the time of need to be laminated in. In the assembly department, the first station is called "open hull." You could walk back to the upholstery department and the cushions for that boat inside and out would be the ones being assembled. Two Takt time cycles later when the deck was set on top of the boat or "mated," the cushions would be wheeled out on a cart and set on top of the deck where they could be pulled in through the hatch when the boat was ready for their installation. This was an unbelievably synchronized schedule. From a Lean standpoint, it was a *push* system. It was not Kanban or pull, but it was as close as a schedule could be.

STANDARD WORK

One thing that Sea Ray practiced was the use of "job duties" or standard work, line balancing, and load leveling. This may not have been the more developed standard work as we have today, but it was standard task lists, and the installation of each component was expected to be standardized or done the same way each time. I think it was awesome that supervisors did this all by hand. When the company set the rate or Takt time, the supervisors went to their desks and created "job duties." Each supervisor knew all the tasks it took to build the boat on his line and how long it took to do them on average. If the Takt time was 2½ hours or 16 boats per week, he gave each boat builder 2½ hours worth of work. He would also give one person a little less to provide slack time. All the materials were delivered from the warehouse to the boat on a kit cart and arranged in an order that complemented the order of assembly or installation. So each boat builder had a job duty list and all the materials he or she needed to do their job. The supervisor would determine the number of stations in the assembly line by dividing the cycle time (time it took to build the boat) by the Takt time (how often). So if it took 15 hours to build the boat complete in the assembly line and they build 16 boats per week, which is a Takt time of 2½ hours, they set up the line with six stations. Make an error in these calculations and there will be trouble.

Albeit that these activities were driven with a push schedule, Sea Ray was practicing these Lean principles before there were any mainstream Lean books and Toyota was anything other than some funny-looking foreign car. On the other hand, Sea Ray rarely ever produced a boat that was not sold to a customer. As far back as I can remember, there was always the name of a customer on the order form on the back of every boat. I think this also served as a constant reminder that there is a customer out there. Although I did not hear him say it personally, people have quoted Connie Ray with the statement that "Nothing happens without an order."

After completing the management training program over the course of a few months, I arrived at the factory fresh out of another hitch in the Army, new MBA, lots of enthusiasm, and passion that bordered on aggression. I was so excited. I was walking on air. I reported to the operations manager and was given a list of people and the guidance that I may want to fire all thirty-six of them and start over. It was a terrible department and it was broken. When I entered the first of two warehouses, I could not see ten feet in front of me. My new boss, the operations manager, was who I was cleaning up after, and also the plant manager was edgy about me being there. He was new and insecure, and I, being hired by his boss, was not "sitting well" with him. He showed me around and turned me loose. The plant was in the "South" and I was from the "North." Major culture shock. I was constantly reminded that I was from the North, and it was challenging to fit in. I did not like that part.

The first thing I did was go to HR and photocopy the ID photos of everyone onto a piece of paper. I carried them on my clipboard and any time I saw someone coming, I would call them by name. I used it until I knew it from memory. It is hard to be a leader when you do not know people's names.

Lessons Learned

- Factories and organizations are like small communities with their own culture.
- Do things right, with a focus on doing them right the first time along and a value that things are not done until they are right is a powerful element in being successful.

- Setting up and planning ahead for work is the fastest way to get work done.
- Standardized work is highly effective, especially in training new people.
- The way new product is released into an organization is a key factor in achieving success.
- Leaders who understand and know how to do the work make great trainers of people.
- Truly understanding quality and its importance as a cornerstone is a huge success factor.

4

My First Process Improvement Project

I met with the team and told them what we would be doing: creating a well-run, best warehouse in the company! I worked day shift and night shift. I arrived before the day shift, said good night to them when they left, greeted the night shift, and left shortly after they left. This was so much fun. I worked so many hours that security searched my car one night at 3 a.m. when I left. They could not believe that any sane person would work that many hours. I must be up to something.

You could not even walk through the warehouse. It was completely unorganized. There were boxes and boxes, and absolutely no rhyme or reason to anything. Production would scream for a part such as toilets or batteries. Purchasing would say they had just ordered them. However, production was in need, so they would order more. A week later, three pallets of brand-new toilets would be discovered in the back behind something. Unbelievable! The stock location system consisted of polling everyone whenever someone needed something to find out if anyone knew where it was. There was a line of thirty-plus people outside the stockroom window to get missing parts from their materials kits. It was a mess.

Early on in my career, I developed a personal approach to turning around processes, departments, or areas. I focus on people, processes, and quality. Generally speaking, I focus on applying 5S and workplace layout almost immediately; and while I do that, I get to know the people. One thing I do is observe who is *really* leading, not just the people in leadership positions. Some of the strongest leaders in factories are not in leadership positions. It is important to know who they are and involve them. After that, I focus on the quality performance of the area and develop a strategy to improve. This all fits in nicely with following Lean principles.

So, I rallied the people to get the place organized. In my office I had a dry-erase board, and I placed a five-by-ten-foot piece of laminated plywood over two filing cabinets for a desk. This was before the use of personal computers proliferated. I used stacks of yellow legal pads just like I use Windows® Explorer® today. One subject per tablet and all the action items I had to manage it. So I had tablet after tablet of things to do.

I took a piece of cardboard and drew a scaled drawing of the main warehouse and cut out scaled racks and equipment from yellow paper. The next day I asked for volunteers to help design the new layout. The team got into it (they were in the early stages of designing their process). We decided to arrange the warehouse by commodity, which was also the way accounts were assigned to buyers. Then we had a Kaizen event. I will bet the guys worked eighteen hours moving all the racks with their forklifts. They placed things to the nearest quarter-inch. They created locations and set them up in the computer. They put part number labels under every part, and they put orange tape around the boxes or parts farthest back on the shelf. Anytime someone opened a box or delivered an item with orange tape on it, they notified purchasing about it. They could then double-check the inventory to avoid a stock-out. I personally waxed the concrete with Future® floor wax. A lot of pride was created that day.

After three months, the inventory was much more accurate, the place was immaculate, the floors in the warehouse were shiny, and the warehouse staff were working as a team. With the warehouse organized by commodity and sub-commodity, we assigned a warehouse person to each buyer (they bought by commodity). I called it "warehouse-purchasing counterpart teams." The buyer and his or her teammates worked together to keep proper stock levels. We were receiving lots of compliments, and people were putting in transfers to join the warehouse team. Almost all of the ideas and actions we took came from the team.

We had a great team with great morale. Being just out of the Army, I offered to run PT with the guys at lunch, and we did…in formation! That is a true story. For some reason, I could not imagine doing that in today's work environment. There was a lot of crazy fun among the employees. One day a female sales representative was in the warehouse. One of the guys "passed gas." She asked what the smell was. One of the guys told her it was the smell of some rubber gaskets she was standing next to. The next thing she was doing was holding up some gaskets and smelling them, and saying that she could not smell anything! Then there was this guy who worked in the assembly department. We had a guy working in the service part shipping area who was hilarious. His name was Joe. He worked with an older guy named Jim from the warranty department. Although Jim was a very nice man, he was a crusty old codger with a temper but you could not help but like him. One day he had a big box he was getting ready to ship out some replacement seats for something for a boat. Joe climbed into the box and covered himself with Styrofoam® popcorn. When Jim went to put something in the box, Joe jumped out of the box and scared Jim half to death! If Jim chewed tobacco, he would have swallowed it! Maybe you had to be there and know these characters, but they were hilarious. They also knew and did their jobs very well.

After organizing the warehouse and putting up all the stock, we identified $100,000-plus in obsolete inventory. When the senior vice president came through, I showed it to him. He said, "You need to start getting rid of that stuff." A couple of months later on another walk-through, I proudly showed him the empty shelves where the obsolete inventory had been. He said, "I thought it would take you more time than that to get rid of it." I said, "Oh no, it was easy; we had it all in the dumpster the day after you left last time." He gave me a look I will never forget. We had a "moment." A moment when I realized that when he said to get rid it, he did not mean that we should throw it away. Lesson learned!

In addition to the main assembly warehouse, we had a large bulk item warehouse in another building where we stored plywood, fiberglass, and engines. Corporate also used this warehouse to keep an engine inventory to supply other plants. I continued to work a lot of hours. One Saturday, I was working and needed to check on something in the bulk warehouse, so I headed over there. When I entered, I was surprised to see a truck with a boat on a trailer behind it and one of our department managers was buffing the boat. I had a fleeting thought in my mind, questioning why he would be buffing a boat in the middle of the warehouse. Anyway, I approached him (he was my teammate) and said something like, "Hey, if you don't mind, can you buff your boat some other place next time? We have a big job improving our inventory, etc., and we are really trying to restrict access to the warehouses." He said, "Sure, no problem."

About a month later, we had a big inventory. A young lady named Kim managed our engine inventory. She said that we were missing two engines. They were 454 Magnum Mercruiser® stern drives. They were big engines. Still being fairly new, I did not know if this was an inventory error, miscount, etc. It would be fairly hard to get an engine out of the plant without someone noticing it. Our plant manager was none too happy about this situation. He took it seriously and, if my memory serves me correctly, involved law enforcement somehow. I kept racking my brain as to how someone could get an engine out of a warehouse that was locked all the time and get it past the guards at the gate.

One night I had just gone to bed and was doing some thinking. All of a sudden, I got this picture of a manager buffing his personal boat in the middle of a warehouse full of engines and a forklift sitting about ten feet away. I'll bet the guards said, "Man, you sure got a shine on that thing," as it pulled out of the gate. You do the math.

LEADERSHIP NIRVANA

By Christmas of 1988 I had been leading the materials team for five months. This was five months of hard work—and I mean hard work. I pushed and inspired the people. They were really gaining pride in their efforts. People from other areas in the plant were applying for transfers into the materials group because they wanted to be part of progress. I was personally gratified beyond all expectations at Christmas when the materials team presented me with a rifle as a gift! In Tennessee when the guys give you a rifle, that is "leadership nirvana"! To me there is nothing more gratifying than having people you lead compliment your efforts with appreciation. That was fun.

One day I approached the plant manager and inquired as to how he thought things were going. He said, "Very well." So I asked him how soon I could expect to be promoted to staff manager. He said, "Not any time soon." I asked him why. He said, "It was never intended to happen that quickly. Maybe in a couple of years." "A couple of years?" I asked. He nodded. Being a lot younger and maybe a little too passionate, I became upset. It was in the morning. I walked out of the factory, got in my car, and drove fifty miles to corporate headquarters. I advised the receptionist that I was there to see the vice president of operations. She said, "Go on back." I said, "Hey Harry, how are you?" He said, "Fine. What brings you to corporate?" I said, "Well, I need to talk to you about something." He said, "Sure." So I asked him, "Do you remember when you told me that if I fixed the factory warehouse, that I would be promoted to staff manager? Well, I fixed it and Henry says I won't be promoted for a couple of years." The VP of operations looked at me and said, "Now wait a minute. You just need to be worried about the fact that you have a job and are earning a fair salary." I said, "You made a commitment to me that if I fixed that warehouse, I would get promoted. That is what you said." He said, "Well, I didn't expect you to work morning, noon, and night and get it done in two months!" I said, "OK, I have said my piece. Thanks for listening." So I got in my car and drove back to the factory. The next day, the plant manager announced that I was promoted to materials control manager, a staff-level position.

About six months later, the plant staff met in our plant conference room and advised that we needed to come up with a plan to significantly increase production. We had discussions and did "what-ifs" for a week. About thirty days later in the same conference room, we were advised that we were laying off nearly the complete labor force and keeping a "nucleus" of management and key employees until things picked up again. Ultimately, I moved on, and the factory was later idled and then sold.

Lessons Learned

- People do love to participate in creating the process they work in and are a powerful source of ideas and solutions if you provide them the opportunity and listen to them.
- People do not mind working very hard, and they will often work unbelievably hard if they are working toward a goal or solution that they participated in creating.
- People do like to be part of positive change if it is led in the right way.
- Politics is everywhere.

5

Process Improvement Project #2

Shortly after leaving Sea Ray®, I received a call from Chris-Craft®. They asked if I was interested in joining them, and I flew down to Florida to interview. They treated me well and I was interested. However, I was thinking that the position was in Florida. I was all jacked up at seeing the palm trees and sunny Florida. To my surprise, they offered me a position as the materials manager in their Swansboro (North Carolina) plant. Oh well; but hey, it was a new adventure and a new opportunity. This plant was originally a Uniflite® plant purchased by Chris-Craft. Most of the employees still had "Uniflite blood" in them. On the other hand, they were very talented boat builders. I noticed right away an accent that I had never heard before. It was called a "high tider" and is a unique dialect among some of the people along the coast of North Carolina. In addition to oyster bars along the coast, bars were filled with people doing a dance called the shag. Looked like a type of sock hop to me, but man did people have fun doing it!

After the heyday of Chris-Craft in the 1960s, the company was passed around and grew much smaller. In the early 1980s, it was owned by a man named Murray and known as Murray-Chris-Craft®. In a move to buy transoms to put their Mercury® outboards on, Brunswick® made the bold move of purchasing Bayliner® and Sea Ray® from their private owners, Orin Edson and Connie Ray. They both became wealthy, and the companies grew throughout the late 1980s and 1990s. In response to this move by Brunswick, OMC purchased several small boat companies. Another part of this industry transition occurred when a shrewd businessman by the name of Irwin Jacobs, the owner of another industry giant Genmar®, participated in the auction and bid up the price that OMC had paid for Chris-Craft considerably over the market value.

This is the beginning of one of the most unbelievable experiences in my career. I moved to Swansboro and joined the Chris-Craft team. It was my first exposure to what I would call "boat builders" as opposed to "boat assemblers," which is a basic, yet very important skill set, in smaller production boat plants. Chris-Craft Swansboro built product from the thirty-four-foot o the fifty-six-foot Chris-Craft Constellation®. They also built the Classic 427 Catalina®, which was a great boat. These boats were simpler and more practical, but really great boats. I began to organize the warehouse, get the inventory right, and teach the buyers how to use MRP, which also meant getting sales to better manage the master-manufacturing schedule. It was going along fairly well, as establishing materials and purchasing processes was a fun challenge.

One situation that popped up was when I received a call from an irate dealer because he had just received a brand-new boat with a brand-new set of engines. The only problem was that the engines were three years old! Seems like the materials crew was practicing "last in, first out." They stacked all the engines in a high stack outside the factory. Unbeknownst to management, each time they got in a new shipment of engines, the materials guys would place them on top of the stack and then when production called for a set of engines, they pulled them right back off the top! This left quite a few underneath the pile to get old. Lesson learned—and a nice discount to the dealer.

THE NEW OWNERS ARE AT THE GATE

1990 was one of those years in the marine industry. Tough market, lots of players, dynamics of the "big boat company consolidations" still taking hold. I continued on as the materials manager and was quite busy. I am sure it is true in all businesses, but the rumor mill in the marine industry used to be quite the thing. If you were going to be in management in the marine industry, you had better develop the ability to ignore the rumor mill. Otherwise, it would drive you nuts. What was even worse was that way too often the rumors were true. It could be quite distracting.

Well, in1990, the rumors that the Swansboro plant was closing began to surface, and the number of rumors increased over time. It started to bother me, so I approached the plant manager. He was an "old-school" manager, kind of an honorary position. Just be in charge, don't rock the boat, and schmooze a little with upper management. He might be seen on the factory floor once a month and it would be at an odd time that most would miss him. He came in at 8:30 a.m., took about an hour and a half for lunch to hit golf balls, and was gone for the day at 4:30 p.m. Although there was financial pressure in the marine industry in 1990, it was nothing like it is today. The difference in financial pressure between 1990 and today can be likened to the difference between a child getting a spanking versus a "Singapore caning." But I digress.

So, I said, "Stan, there sure are a lot of rumors out there about us closing." He said, "Come on Bill, you've been around long enough to know better than to listen to the rumor mill." I asked, "Are you sure? Because there are a lot of rumors." He said, "There is no truth to these rumors; you need to go back to work. If I hear anything, I'll let you know." So I went back to work. He really had no clue.

Being the motivated young man I was as a young bachelor, I worked a ton of hours. I say worked; but to this day, I often say I have never worked a day in my career. I am one of those people who have been lucky enough to find a way to get paid to practice my passion. Since I was a young man, for what reason I do not know, when my feet hit a factory floor, I get a burst of energy and passion from inside that to this day is unending.

Well, one Sunday afternoon, I was in my office working away when the security guard, Sgt. Gonzales, (a proud retired Marine) knocked on my door. He said, "Mr. Trudell, the new owners are here and they want to look around." I said, "What?" He replied, "There are two guys out here and they say they are the 'new owners.'" I am saying to myself…OK…this is going to be interesting. So I went out and introduced myself. I said, "Hi, I'm Bill, can I help you?" A gentleman named Will said, "Yeah, you don't know it yet, but we're taking over this place, and we are here to look at real estate and thought we would stop by." I was thinking this was unbelievable. This situation was surreal. I asked, "Do you by chance have a business card?" He gave me a card. It was an OMC card, and said something about some boat group…I can't remember the specifics. So I said, "Well, make yourself at home."

On a side note, this guy Will had an arrogance about him. Actually, to me, he was just plain weird. He actually had a dog on a leash, and away he went into the plant. He carried himself like he was some Roman conqueror or something. I was not impressed. The other guy who was with him was high energy and somewhat more reserved. I also noticed what looked like some family or people waiting in their cars. They walked throughout the plant for about an hour. I heard a couple of comments about what went where, etc. They had nothing else to say to me, and a while later they drove off. They were not friendly. There were no "Thanks for letting us in" or anything like that.

Monday morning I came into work bright and early as usual. I was somewhat tense. Plant closings and big management changes were beginning to happen more and more across the industry, and I really did not want to become unemployed. The previous plant I had worked at was closed. Anyway, when Stan got into work, I went up to his office and said good morning. I said, "Stan, are you sure there is nothing to the rumors?" He kind of growled at me and said, "Bill, I thought we talked about the rumors." I said, "OK, but some guys came in here yesterday and said they were the new owners, and they gave me this business card." I could see the blood rush from his face. He said, "I'm calling Florida, and I'll let you know. Keep this between you and me."

My office was located just off the front lobby. I had a direct view of our receptionist Beth from my desk. She was a bright young lady, great personality, and very good looking. I really liked Beth. Well, it was Tuesday morning and I had not heard a thing from our general manager. About 9:30 a.m., I heard Beth say, "Oh oh, that isn't good." I asked what happened. She said, "Barry and John (HR and president, respectively) are getting out of a car in the parking lot and we are not expecting them." "That can't be good." Next, Mack, our HR manager, steps out of his office and says, "I'll tell you what I think. I think the fat lady is getting ready to sing." Drama at high noon. I just laughed out load as I am writing this. This was crazy.

Our president came into the lobby. He was a very classy and professional man. He acted how you think a president should act. He was not only the president of Chris-Craft, but also the president of the OMC recreational boat group. On the other hand, the director of HR he had with him was the proverbial "snake" that makes people not like HR. (More on him later.) Once they were in the lobby, Narley turned to the right and went right into our HR manager's office and closed the door. John turned to the left and came into my office. He then closed the door and proceeded to say, "Bill, we need to make some changes today, and I am going to need your help. Can I count on you?" I said, "Sure John, whatever I can do." He said, "I have to talk to Stan, and then I will be back down to talk with you about the help I need." I said, "OK."

About forty-five minutes later, John came into my office and closed the door again. He said, "Bill, we're closing down this plant. OMC is going to use it to build some other brands—Hydra-Sport®, Donzi®, and Topaz®. They already owned Hydra-Sport and Donzi but evidently, as part of this effort, had purchased some fish boat tooling and brand name from a company called Topaz. I just let Stan go, and I would like you to be the plant manager of this place and manage the transition out of here and when you get done, I will have a job for you in Florida." Being the person I was, I asked John, "What kind of job?" I had engaged John about being a plant manager on more than one occasion. One time when I was pressing him about being a plant manager, he mentioned that I had plenty of time and that I might need to mature a little more. He pointed out that I was only thirty years old. I remember telling him at the time that a guy named Bill Clinton who was thirty-three years old was the governor of Arkansas; and if he could run a state at thirty-three years old, then I could run a factory at thirty. John gave me a stern look and said, "Bill, you need to learn patience." John and Barry departed. As they left, the word "Patience" on a placard I had put on the back of the nameplate I had on my desk caught my eye. John was right. I was not ready—yet. I was prepared, but I was not ready.

I went to lunch with John and Barry, and we decided who would stay and who would go. There was no need to keep on some people with this change in direction. After we came back from lunch, Narley and Mac (the plant HR manager) laid off the people to be let go that day.

We then called a company meeting in the lunchroom. I explained to the employees that we were closing down Chris-Craft operations and that it appeared that OMC was moving three other brands into the plant. I told them that that was all I knew. It was tense. Employees started asking if they were going to have jobs, and they were not happy. What does this mean? What is going to happen? What about benefits? All I could do was assure them that I would communicate to them any and all information given to me as fast as I could. I also committed to the boat builders that I would personally write a letter of recommendation for each and every one of them and hand-carry them to the folks from the new group. This was a commitment I kept. That meeting was a growth point for me as a professional. It was tough.

Over the next few months, I was involved in the layoff of nearly 200 employees. It was not fun. The people of the marine industry mean a lot to me, and I develop very close relationships with boat builders. I believe that you cannot lead those you do not know. So this was tough. I gave pink slips, handshakes, and hugs to a lot of people. This was one of the most challenging situations that I had ever encountered.

The company asked that I clean out the general manager's office and dispose of any company documents, etc. So, one morning I headed upstairs and went in an office. It is always a little strange to go into an office of someone who is no longer there, yet it is interesting. So I sat at the desk and started with the in-box; it was full. I kid you not; there were memoranda in his box that were more than three years old. There was dust on everything. I opened one of the desk drawers. There were a couple of newspapers and a *Playboy* magazine. There were files and files of papers everywhere, and they were the insignificant kind that should have been thrown away—again, three, four, five years old. I could not possibly imagine the office of a manufacturing manager looking that way today. My desk usually has a laptop, a couple of Lean books, and maybe a small stack of paper that others have given me, and they disappear quickly. "Times, they were a-changing," as Bob Dylan sang.

During this time, the management from Hydra-Sport, Donzi, and Topaz arrived. They brought an HR manager and began to hire. These folks were all on a political power trip. They were arrogant and flat-out disrespectful to everyone. Both of the key managers bought huge luxury homes and made sure everyone knew about it. I could not wait to get out of there. One day we were boxing up hand tools, air tools, etc. to ship to Florida. I was called into the office and dressed down for doing what I had been specifically directed to do by the management at Chris-Craft. So there I was in the middle of a political pissing contest between vice presidents, presidents, group presidents...over a wooden crate of tools. I hate politics.

So, I managed the transition of the plant. I arranged to move inventory, shipped tooling, moved obsolete tooling to the fields, packed up documentation. What was fun, however, was that I led a group of boat builders in building the last two Chris-Craft 501 Constellation® motor yachts and the last 392 Commander®. The Constellations were fifty-six feet long and the 392 was thirty-nine feet long. One was going to Japan and one was going to Taiwan. These were both classic yachts. They were beautiful boats with old-style yacht styling and classic boat building by very talented boat builders. There was some real boat building talent on the eastern shore of North Carolina. I miss some of those people to this day. Although nothing lavish or big, all the offices in the plant had some very stylish office desks and furniture. Rich woods and fine construction. I realized one day that the boat builders at the plant had built all of them. Amazing!

There was a certain situation that occurred in the plant that I often think of any time I am pondering the subject of leading people through change and human nature. I wish I had a picture because it would be worth a thousand words, at least to those of us who lead change. Craftsmen at the plant used routers, saws, and other cutting devices to cut the hundreds of wooden components required to build the boat. To speed up the process, they used patterns made of wood. They stored them by hanging them from the ceiling with hooks, stacking them in places, keeping them on carts. There were probably 300 or more patterns in the wood mill. Here is the unique situation. In the middle of the mill was a CNC router with about four patterns stacked on the cutting platen! The company probably spent $100K to buy a CNC that would eliminate the need for patterns, reduce scrap, and easily store files for replacement parts. The boat builders said it did not work. Most of the systems I have implemented in companies over the years "didn't work" when I got there. Obviously, some change leadership was missing.

After six months of laying off great people and dealing with the upper-level politics between Chris-Craft and the staff of these new brands moving in, I was tired and stressed. Frankly, I was exhausted. As things wrapped up, my last task was to accompany the two 501s down the Intracoastal Waterway to Charleston, South Carolina, to the port where they would be loaded onto a freighter and delivered to their destinations. We had a couple of boat builders who were also captains and who plotted our course. One of them was nice enough to let me take the helm for hours at a time down the Intracoastal. It was so cathartic to sit up on the bridge of a sixty-foot motor yacht on a beautiful waterway. Nothing but blue sky, smooth waters, and peace. The remnants of Hurricane Hugo were still obvious. There were downed trees and debris everywhere. This was also my first experience at learning that twelve inches of water and twelve feet of water look the same. I was reading charts, which indicated that the water on the port side was a foot deep. However, for some reason, it looked much deeper to me. A little mud got kicked up, and the charts got used the rest of the way.

Coming into Charleston was kind of neat. I noticed a fort-looking structure on a small island. It was Fort Sumter. How cool! We motored right by it. We settled in Charleston for the night. We had to deliver the boats to the port the next morning. That night, we went downtown to the historic district and went to a great steakhouse. The next morning we went to the port. The two boats were docked side by side, awaiting being hoisted via crane onto the freighter. As I was looking at them, something caught my eye. They were supposed to be identical boats, but something was a "bit off." When windshield windows are installed on a large boat, they are inserted into holes cut into the fiberglass deck. The windows had a six-inch border of black paint around them for cosmetic effect, at least on one of the boats! The other boat did not have those borders. Actually, it did not look that bad. We had to make hoist time at the port, and the factory that made the boats was now closed and had no employees. But to this day, I wonder if anyone ever noticed. One boat went to Taiwan and the other to Japan. Somehow I doubt it.

I delivered the boats, packed up my belongings, and headed down the road to Florida. I was stressed and tired—very tired. While driving to Florida, I stayed the night in a hotel along I-95. When I got up the next morning, I unknowingly turned back north instead of south and drove almost three hours in the wrong direction! Oh well....

After moving to Florida and beginning to work at Chris-Craft, we continued to hear of the activities in Swansboro. Many of the employees were re-hired by the new organization. OMC had awarded them quite a bit of money (millions) for capital improvements. They began construction on a state-of-the-art fiberglass lamination facility. It was incredible at the time. It had a state-of-the-art air make-up system with stacks and ventilation to create a good working environment and meet all the air quality regulations, etc. The Swansboro site was quite attractive, nestled on the Intracoastal waterway, so that made it great for product testing and water delivery of boats. In addition to the lamination building, they also began construction on a multi-story office building overlooking the water. This was going to house engineering and administration. This was a beautiful building.

So, after being laid off and not knowing whether or not they were going to have a job, these were exciting times for the boat builders in Swansboro. Actually, it was exciting for the whole community. The population of Swansboro was right at about 1,000 in 1990. When a Fortune 500 company builds a factory and announces lots of jobs in a community of that size, it is a huge thing.

Well, get ready to file something under the "unbelievable" tab.... (I did not witness what I am describing. This is how it was described to me.) So the office building was done, and the new lamination building was nearing completion. This description of these events may vary from reality somewhat, but after listening to many accounts, they are very close if not right on. The company scheduled a big celebration event for the new lamination building. The CEO of OMC was coming to the plant to attend the celebration himself.

On the day of the celebration, the place was abuzz. Will and Chip, the executives in charge (the two guys who showed up as the new owners on that Sunday afternoon and bought luxury homes) headed for the airport in Jacksonville, North Carolina, to pick up the OMC CEO and bring him back to the plant for the christening. All the employees were scheduled to be in the lunchroom upon his arrival for a speech and start of the celebration. When the executives arrived at the airport, they walked out to the OMC corporate jet. (Now this is the story that I have heard from more than one source, and I believe to be true.) The door opened and the CEO and the corporate attorney came down the steps of the jet. The execs greeted them with enthusiasm. Then they noticed that the CEO was not smiling—and neither was the attorney. The CEO advised them that the decision had been made to shut down the facility. The attorney was there to explain the details. Can you possibly imagine what was going through the execs' minds?

A few moments later, Will and Chip were headed back to the Swansboro plant. They were proud owners of two luxury homes in the middle of nowhere, had relocated their families, and were headed back to a lunchroom full of employees expecting to hear an exciting speech from the corporate CEO about moving onward and upward. The execs had to go into that lunchroom and tell the employees that the plant would be closing within ninety days and they were all losing their jobs. I cannot imagine what that must have been like. This was one of the signs of OMC's impending demise, which ultimately ended in bankruptcy. Actually, a very sad story about a great company with unbelievable product (Johnson® and Evinrude® outboard motors).

Lessons Learned

- How much rumors can impact an organization by sapping energy and clouding focus.
- Excessive politics displaces leadership, teamwork, and ultimately signals demise as, by nature, it focuses key people on political agendas versus the strategic direction of the company.
- Building empires is not the same as building a great company, and these are mutually exclusive.
- Bad leaders exist at all levels.
- The construction of a new facility in no way ensures a bright future.

6

My Lean Journey Begins

Shortly after completing my assignment in Swansboro, I arrived in Bradenton, Florida. It was November. I had never spent much time in Florida other than to interview with Chris-Craft and a few spring break visits. The weather was beautiful, and people were really friendly. The thing about Florida that is fun is that starting in September, the weather gets nicer each month until late spring when it gets hot again, unlike "up north," where the weather gets worse and worse.

I showed up at Chris-Craft and they made me the materials manager. The company had inventory everywhere—and I mean everywhere. They had hundreds of thousands of dollars of obsolete inventory. They had an MRP system, but were not using it. The BOMs (bills of material) were inaccurate and missing key components such as plywood, fiberglass, or resin. The perpetual inventory was inaccurate. People walked in and out of the warehouse like it was a hardware store. Production was screaming about materials shortages, and the stock-out list was a mile long. Because the BOMs were inaccurate, the pick lists were inaccurate. So if the kit for one boat did not have a bilge pump in it, a boat builder would swipe one from another boat's kit. Then when another boat builder went to install the bilge pump, it would not be there and the materials team would be accused of missing a part on that kit. Holy smokes! What a circus! Delays and downtime were occurring daily. Honestly, it was not necessarily that different from most other boat companies in the 1990s. With all that said, it was a fun atmosphere. Florida boat builders are a unique group and very talented. All kinds of nationalities and languages, and for the most part everyone seems to get along. Over the years, I have grown a tremendous appreciation for their talent.

Also, upon arrival, management said, "The purchasing manager doesn't know what he is doing so you will probably need to get rid of him right away." My response was, as it would be today, "If you want him gone, then you fire him now; otherwise, everyone always gets at least ninety days to prove themselves to me." They did not fire him, and neither did I. He turned out to be one of the best employees I ever worked with. A few years later, I got a call that Lazaro had died of a massive heart attack at thirty-nine years of age. He was having bad indigestion and thought it was a side effect of a medication he was taking. He was wrong and fell over dead while making breakfast. Lazaro was an awesome human being. He would sometimes talk about coming from Cuba at five or six years old; he took off his shoes before getting on the plane and gave them to his cousin. What an incredible person Laz was, and an awesome worker and great teammate. May he rest in peace.

At this stage in my career, my exposure to Lean manufacturing had been minimal. But I did know what I knew and knew what I did not know, and I was smart enough to ask for advice about what I did not know. To this day, I see even senior executives who do not know what they do not know. I mean, knowing what you do not know is just as important as knowing what you do know, right? I gathered people in rooms and asked for advice on what the problems were. I listened to upper management and anyone who would want to talk to me.

So, here were the problems:

- Production was interrupted due to high stock-outs.
- Thousands of dollars of obsolete inventory everywhere was coming right off the bottom line.
- Large WIP (work-in-process) inventory shrinks and variances totaling thousands of dollars were occurring.

Root causes and other:

- The buyers would not use the MRP system.
- The product BOMs were inaccurate and improperly structured.
- The MRP output and reports were inaccurate.
- No one understood how the MRP worked.
- Inaccurate inventory.
- The engine inventory was high and experiencing obsolescence (this was a problem because new boat owners expect new engines)

So I went to work. I looked at turning around areas like solving a Rubik's Cube. It is just part of my nature to enjoy solving process problems. The interaction with people during the problem-solving process is also very enjoyable. People almost always have the answers if you have the right questions. You know, some people say if you want to get to know your kids, play with them. Well, I say if you want to get to know your employees, improve processes with them.

The buyers would not use the MRP reports because nothing was accurate in them. Materials requirements—inaccurate. On-hand inventory—inaccurate. Ending available inventory—inaccurate. We locked down access to the warehouse and took daily cycle counts. This got the inventory accurate.

Over time, the buyers started using the MRP report. Getting buyers to transition from manual purchasing practices to using an MRP output is quite a challenge from a change management perspective. Buyers "hang on tight." There is a lot of pressure to avoid stock-outs; and when there are stock-outs, it is unpleasant. With that said, getting buyers to trust and believe in an MRP report that has been inaccurate in the past is a challenge. So, it takes leadership, training, and coaching, and reinforcement of new work practices.

Stock-outs were still occurring because the MRP requirements were inaccurate from the BOMs being wrong. I went to the plant manager, who was a good guy, and said, "Tom, if we don't get the BOMs right, you can't blame me for being out of materials." He responded, "What do you want me to do, move the BOM management from engineering to materials?" I said, "Yes," and he did. We moved the BOM management responsibilities into the materials group and gave our BOM technician direction. (Engineering threw a fit.) He set up the BOM by product structure, and we gave him support in auditing the product.

We worked hard to get the inventory and the BOMs accurate. And I mean very hard. Hours and hours of hard work, communication, information gathering, BOM maintenance, etc. Our BOM technician did a great job setting up indented BOMs following product structure formatting. With proper formatting, anyone could easily find information in the BOM, or root cause problems. However, even with all this great effort, the stock-outs and disruptions to production continued. I was becoming stressed. It just seemed like we could not win. We did find that production was just throwing away damaged parts and requesting replacements that would cause stock-outs. We fixed that. I had now thrown "everything I had" at the materials function and it did not seem to be getting any better. I have been blessed with confidence as a person. I do not really know why, but I have always been confident in my abilities. Well, my abilities appeared to be limited given the meager results we were getting. I was actually wondering if I was going to fail, and boat companies did not think anything of whacking a materials manager and replacing him or her. The fear of failure was not fun.

Then over the course of a few weeks, stock-outs began to decrease more and more, and so did the inventory. Our inventory turns were going higher and higher, and stock-outs were going lower and lower. We actually had a day with "0" stock-outs! It was a huge relief and a fun win for all the people on the materials team. They were holding their heads high with a sense of pride, and production was saying, "Good job." I got a note from the plant manager congratulating the team.

So things were rolling along quite nicely in the materials group. It was the end of the year and time for the dreaded inventory. Outboard Marine Corporation (OMC) always sent in Arthur Andersen auditors to do test counts, WIP maps, missing ticket reports, the whole ball of wax. The inventory took a whole day plus into the evening. No one could go home until all areas were test counted and cleared by the auditors and the missing ticket report showed no missing tickets. Employees became quite testy, wanting to go home before the end. But it did always end. One time, several employees left without permission and went down to the local watering hole—scallywags. I sent someone after them. When they returned, they were inebriated. I had to keep them until they had sobered up. Hey, what are you gonna do?

We completed the inventory and it showed an unfavorable inventory shrink of somewhere around $750K. The CFO was angry. He was not a happy camper and immediately began to blame the materials group. I did not understand that at all. I could not relate to anything we had done that could cause a shrink. I apologize for stating the obvious, but a *shrink* occurs when the computer thinks you have a certain amount of inventory in a category, but when counted there is a whole lot less, which results in an immediate write-off against the bottom line. Not good.

So the American "manufacturing blame game" started. Our CFO was aggravated and was blaming the materials group for the shrink. So I started talking to people, studying my accounting books from business school, and doing some analysis. I had to find the root cause. This was a very political environment and, unless someone could find the cause words, blame and posturing would ensue.

Chris-Craft used a standard accounting system and managed "standard boat bills of material" in accounting. Accounting would create a standard

BOM on a PC. They would only update or roll the standard into the MRP BOMs twice per year. So for a few moments twice a year, manufacturing and accounting were using the same BOMs. Manufacturing used the BOM function in the MRP system…multi-level, product structured, etc. As with any MRP system, we would create jobs in the Master manufacturing schedule, which created MRP demand, etc. When a boat was ready to start, we would "allocate" material and create a pick list. The pick list items were picked and issued to a boat in work-in-process (WIP). When the boat was complete, the materials were back-flushed out of WIP into finished goods until invoiced.

The problem was that for some reason, accounting disregarded the back-flush from WIP. Whenever a boat was completed, accounting would remove a "standard BOM's worth" of material from WIP into finished goods. On top of that, the entire accounting standard BOM did not reflect all the BOM transactions resulting from six months' worth of engineering change notices. By the end of six months of this, the values of the MRP BOM used to issue materials into WIP was much higher than that of the accounting standard BOM on their PC. An average difference of $1,000 between a boat's MRP BOM and its accounting standard BOM per boat across six months' production, which could be 700 boats, could easily result in an inventory shrink of $700,000.

I went to our IT (information technology) guy for help. I asked him if he could set up a program that accumulates a continuous list of boats completed, and set up a program that at the end of each month relieves WIP by the amount put into WIP with the MRP BOM. We worked together and created the "WIP Relief" report that was used from then on in accounting. Our CFO was miffed when we briefed him on the root cause and the solution. I guess it made him look bad. Instead of being glad that we solved a huge problem, he was miffed.

Obsolete inventory was a big deal at Chris-Craft. Like most other American companies, they would forecast their sales, load the master manufacturing schedule, and start purchasing. It was a guessing game. Model year changes drove a lot of inventory. We had one VP of sales who insisted on something like seven colors for the year: red, black, blue, green, silver, white, yellow. So, that meant we had to stock cockpit carpet in seven colors, 34-ounce interior carpet in seven colors, and 42-ounce interior carpet in seven colors. One year, that VP created a color scheme we affectionately called the "bumblebee boat." It was yellow and black. The problem was that we had to pre-order vinyl, canvas materials, carpet, and other items with those color schemes. We ordered $20,000+ in stock to support the scheme. Not one boat sold with those colors. Here comes the marine surplus buyer. Oh yeah, and another scowl from the CFO for obsolete?

Like most boat companies, we had to forecast engines for the year. They had long lead-times and you got what you ordered when it was due; you got nothing more and nothing less. Sales never took this responsibility seriously. Getting them to sit still for a minute and help develop an engine forecast was like pulling teeth. They would wing numbers at us and we would place orders. Then at the end of the year, they were forcing engines on customers that they really did not want, just because we had to burn up inventory. Some engines never did get installed and were sold as obsolete to the public.

It was about this time that the announcement that OMC was closing Swansboro was made. I was asked to travel to the Swansboro site to dispose of tooling and other equipment. I saw the new lamination building. A brand-new, state-of-the-art lamination building that had not had a boat laminated in it. Unbelievable! To this day, on occasion this situation blinks into my head and I get a thought of disbelief.

Every day at 9:00 a.m. sharp, we had a production meeting. As the materials manager, my role was primarily to inform the team of the status of any stock-out items, answer any materials-related questions, and respond later if necessary. This was a good group of guys, but sometimes they pushed it by blaming materials for shortages. Sometimes they would say that we were out of something such as hatches. I always had an inventory report with me. I would challenge them with, "According to our report, we have eight on hand." Their response was, "Well, guess your report isn't right then." I would step out and call the stockroom supervisor on the radio. A few minutes later there would be a knock on the door and somebody from the stockroom would walk in and set three or four of the hatches that "we were out of" on the table. Fun stuff. I loved my "production brothers."

For a period of time, the production supervisors groused in meetings about this problem, that problem, and issues in their way while they were trying to get their jobs done, which was to complete boats to the production schedule and make sure the quality was right. They were stressed. They were also expected to achieve high labor efficiencies (hours paid versus standard hours earned). Week in and week out, my teammates felt they were fighting a losing battle and, no matter what they did, it did not work. Our plant manager was growling at them for not making schedule and for their low labor efficiencies. Everyone was trying to do what they had always done better than ever, and it was not working.

The things I learned as a boat builder and a manager at Sea Ray gave me a different perspective, and I wanted to help the guys with it. I challenged them to let me help. Just give me a few weeks and some effort. I convinced them to do two things: put two yellow tablets on each boat in the assembly department and get some orange trucker flags.

We asked the boat builders to take one tablet and write down all the tasks they have to accomplish to get the boat done in their area, and how long it takes to do each task. The Takt time for the line was four hours, or two boats per day off that line. Once we got the tablets back, we added up the times for all the tasks and determined that the critical path for the boat was sixteen hours. This took some discussions with the station leads and boat builders. Their knowledge of how the product came together was valuable in this exercise.

The total hours needed to accomplish all tasks came to eighty-six hours. Ten cycles per week generated a labor demand of 860 hours. Available hours per week per employee was 38, so 860/38 equated to a required staffing level of 23 people on the line. Dividing the critical path of sixteen hours by the Takt time of four hours yielded that we needed four WIP stations in the assembly line (actually five, because the deck and hull were separate parts in the first station).

WATER SPIDERS

With that information, we headed into the conference room and wrote on a sticky note each task and how long it took to do it. We broke them up into four-hour standard work packages. During this process, supervisors brought in their leaders and some of their boat builders for input. They jumped right in with their suggestions about what could be done in what order, and what needed to be done in which station. The supervisor and team leaders dispersed the standard work packages across the WIP stations and then assigned them to the staff. In some places, they discovered they did not have enough people, and in others they had too many. They used this exercise to redistribute the workload and leveled out the line. The good news was that the current staff on the line was twenty-five, so there was an excess of two people. We deployed them as "water spiders" on the line. They traveled up and down the line all day, accomplishing anything they could to keep the boat builders working and keep them from having to leave the boats to go get anything. This was fun and really enjoyable.

On the second tablet, we asked the boat builders to write down any problems that were slowing them down or disrupting them, and keep the tablet on the back of the boat. We then asked the supervisor to go up and down the line all day, checking all the tablets and work to resolve or coordinate getting the issues resolved. We also alerted manufacturing engineering for added support in this effort. This effort greatly inspired the boat builders, as they saw leadership making a concerted effort to help them. It is hard not to appreciate it when people are trying to help you.

By this time I had read the goal and was beginning to read more and more about Just in Time (JIT). My Lean journey was beginning. I had begun to read magazine articles and a book or two. In one article or book I had read about the Japanese using Andon or signals to communicate a problem. It described some plants that had miniature red flashing lights such as on a police car. They placed them on poles in the plant; and any time an operator experienced a line stoppage, they would turn on the flasher. Any manager or leader who saw it immediately called in the other managers on the radio and they worked as a team to solve the problem and stayed involved until it was resolved. I could not find any lights right away, so in addition to the tablets on the boat, I gave each station an orange "trucker flag." These are the flags truckers put on long or wide loads.

OK, so the tablets were in place and the flags were on the boat. All managers and supervisors knew that if there was a problem, they would be alerted on the radio to come and help solve the problem. We were all pumped. About an hour after we started, the assembly department looked like a University of Tennessee football game! All you could see was orange. We were quite busy and I do not think we ever killed all the flags, but we did get several. The schedule attainment and labor efficiencies showed great improvement over the next few months. You know what? This exercise was also a real morale builder. What it really did too was focus everyone on the right thing, keeping the line moving and getting things right.

The on-time schedule delivery for this line improved by over 30 percent. Labor efficiency to standard increased significantly. Quality improved too. Boat building is light assembly and being behind, hurrying to catch up, and fatigue from overtime all take a toll on the quality defect rate. This was a success story; and over the next two months, we deployed the same strategy on the remaining lines in the assembly operation.

Quick story. Our plant manager was unique back then. Heck, they all were. He was a "seminar guy." A couple of times a year he would go to some management seminar on the latest craze. Whatever the subject was, he came back up all fired up and we were his guinea pigs, and God help anyone who disagreed with the "new direction." One time, he came from a quality seminar and announced, "We're going Six Sigma." And we are thinking, what in the heck is that? So we asked him, "What the heck is Six Sigma?" He said, "It's what Motorola does for quality. It is 3.2 defects per million opportunities." So we all went out to the plant and had it in our minds that we were going drive our defects down to 3.2 defects for every million opportunities. The only problem was that the only tool we had was to try harder to do better and inspect more. Today, I am a Six Sigma Black Belt. I have to laugh. I do not ever remember hearing him say anything about the design, measure, analyze, improve, and control (DMAIC) process or reducing variation, or statistics. But hey, his heart was in the right place—just kind of amusing as I look back on it That was the last experience I had with Six Sigma…until I earned my Black Belt eight years ago.

Our plant manager was also a passionate fellow. Actually a good guy, but boy did he have a hair trigger. He had an astigmatism that bothered him regularly. He was really popping off more and more. I made a comment to one of my teammates about it. He said, "It's his medicine; it makes him edgy." Then he tells me to just watch him when he is in the staff meeting. He said he always leans back in his chair shortly before he goes off. He said, "Just watch him." So, a couple of meetings later, he was a little edgy. My teammate kicked me under the table, and sure enough, he was starting to lean back and the next thing you know he was going off on someone! I busted out laughing, and he's says, "What the hell's so funny?" To this day, I laugh when I think about that. I am laughing right now as I write this. As I said, he was a good guy.

THE BUILDING IS ON FIRE

I do not know what it is about Sunday afternoons and me. Once again, I was working on a Sunday afternoon when I should have been at the beach. I went down to the copy room to fax some purchase orders. There was an entrance door that came into the copy room. Right outside was a locked gate that allowed trucks, etc., to enter when needed. Next to the plant, we had a test pond for running boats in, and the company let the local fire departments test their pumps or something in the pond. So, it was not unusual to see fire trucks outside the building. Well, this particular day was different. While I was faxing a document, I heard a commotion outside the door. Clanking and noises were coming from right outside the door, so I opened it to see what was going on. To my surprise, I saw a fireman with a big axe trying to bust the chain. He was really swinging that axe. So I asked, "Is there something going on? Hey, I have a key if you need to get in." He said, "The building's on fire! It's going to blow! You better get out of there now!" He was motioning toward the sky. "You better get out of that building now," he said. When I looked up, I saw an ominous plume of black smoke. Oh, my God! I flew back into the office and dialed the plant manager's number. My hands were shaking so hard, it took me three tries to dial the number. I then grabbed my stuff and exited the building. By this time there were news helicopters circling over the building. There were fire trucks and firemen everywhere. Several were lying out with heat exhaustion. What a sight! The hazardous response team had some technical truck set up with sensors spinning. There were cars and people everywhere.

They put out the fire. Someone had left a resin gun laying on top of a drum in the lamination department. The trigger was leaning against the edge of the drum and it had squirted a big puddle of catalyzed resin on the floor. Heat built up and it caught fire. What a day! The company was lucky. Even with all that smoke and craziness, the damage was minimal, and we were able to resume production after a couple of days. The company was very lucky as this situation could have been catastrophic.

HAZARDOUS SPILL

The fire put everyone on edge. We had special training, extra safety meetings, etc., over the next few weeks. Being the materials manager, I was a point of contact for any kind of hazardous materials incident. So, about five months later at about 1:00 a.m. in the morning, I received a call from the plant security guards that there had been a hazardous materials spill, the county hazardous response team had been called, and I needed to get to the plant right away. I jumped in my car and was driving quite fast down Highway 41 in Bradenton. There was a red light with a police car stopped at it. I blew right by him and through the red light. Well, lights and siren were engaged. I pulled over and a ticked-off cop came to my window. I said, "Officer, there is a hazardous incident at Chris-Craft and I have been called to get there right away." He said, "I am going to follow you there, and you better hope there is; because if there isn't, you're going to jail."

I arrived at the plant and saw the same hazardous operations vehicle that had been there during the fire. I asked the guard what happened. He said, "One of your security guards went into the upholstery department and smelled a chemical that must have spilled." I was wracking my brain, trying to think of what chemicals were being used in the upholstery department. I had worked in a boat plant for a long time and could not think of anything other than some contact cement. I told the firemen that I did not think there were any chemicals in the room to be spilled. I then asked the security guard what happened. He said that when he went into the area, he could smell a strong chemical smell. So a fireman and I put on masks and headed into the area. We could not find anything. We took off the masks. Nothing. We asked the security guard to come in. We asked him if he still smelled it. He said, "Yes." What he smelled was fresh vinyl upholstery, just like the smell of a new car. He was a new guard, and it was his first week on the job. We all went home.

Lessons Learned

- Bad processes—or the lack thereof—can make incredible employees look like they are failures. Take heed, leaders!
- Stay the course. A situation can look incredibly hopeless right before a huge breakthrough. This is especially true in tough situations. As an improvement leader, you must stick to the plan. This can be incredibly challenging as others, including top management, may doubt you, results are not happening, and it can get tense. If your plan is correct and you abandon it at this point, you have a big chance of failure. Stay the course!
- Tablets, pencils, and taking the time to talk with people are more powerful than any PC.
- Assisting people in solving problems is an important leadership responsibility.

7

Outboard Marine Corporation — Marine Power Products Group

My big career goal early on was to be plant manager of a boat factory. Not to sound sappy, but it really was a dream. As a young boat builder, I noticed that the plant manager had what I thought was a "cool" job. I mean, to run a factory full of the hustle and bustle of building boats. Making decisions, solving problems, leading people. All those things really appealed to me. I did not know why, they just did. I also noticed that the plant manager always got a "company boat" with his job. I thought that was awesome. Another alluring thing about the job was that the plant manager usually attended all boat shows with some members of his staff. They would go to the Chicago, New York, and Miami boat shows to look at competitors' product for assembly or design features that they could copy and apply to their product to make them build easier. Tablets and cameras in hand, boat by boat, they would go through and take notes. As I look back, this activity was a little hokey. Now we look to our Lean tools to improve product. There was also a great social aspect to boat shows. It was always a great opportunity to see old friends—almost like a homecoming. When times were good in the 1970s, suppliers rented hospitality suites in the big hotels for customers. Suite after suite of great food and drinks. What a party, and so much fun!

One February, some of our management team went to the Miami International Boat Show to review product, talk to suppliers, and check on sourcing opportunities. This show was—and still is—a very exciting venue. There is always a lot of buzz and activity, not to mention the excitement of boats. One of the bigger supplier representative groups provided an outdoor seafood buffet and open bar at a place called the Rusty Pelican. It was really nice. One of the companies they represented was Tele-Flex®, which supplied control cables. I struck up a conversation with another young gentleman who was from the Tele-Flex company. His name was Jeffrey Black. He noticed my passion for manufacturing while we were talking. He asked me if I had ever read *The Goal*. When I told him no, he immediately said he would send me a copy, and that I would really like it. I still have that book. I have read it more than once and, yes, I enjoyed it very much. Mr. Black retired just recently as the CEO of Tele-Flex after a very long and successful career. I have to think that his appreciation for Lean played a role in his success.

So that was my goal: to be a plant manager of a boat factory. After several years in the marine industry, I seemed to always have missed it by one. While working at Chris-Craft boats in the 1990s, I got a call from the purchasing manager (Ed) for the Marine power products division of Outboard Marine Corporation. OMC was our holding company, and this was during the time I was the materials manager for Chris-Craft in Florida. Ed advised me that he would be in Sarasota in a couple of weeks and would like a tour of the plant. I said, "No problem," and was excited that he had called me for such a thing.

So Ed showed up at the plant, we shook hands and began the tour. In the assembly building, there was a raised catwalk for the purpose of plant tours. This plant was originally built by Donzi® and designed to aid in selling boats. Everyone likes a boat tour. There are also offices along the catwalk for supervision and management. Well, that day happened to be a "black Friday." A rather large layoff was taking place. So as I was giving the tour, I could not help but see over my shoulder that some of my teammates were packing up their offices with grim looks on their faces. This tour was supposed to be exciting. I was actually wondering if I might be called in. That wouldn't be good…right in the middle of giving a corporate tour!

So at the end of the tour, the purchasing manager invited me to dinner that night. Over dinner, he began to tell me that OMC had established a personal water craft (PWC) team. It was a cross-functional team with members from marketing, sales, engineering, manufacturing, and purchasing. He went on to tell me that they had not filled the slot for purchasing because they did not have the right person at OMC. He said that they saw a great parallel between boat construction and that of a PWC, and that further they would share a portion of the same supply base. He seemed to be giving me a pretty in-depth view of this project over dinner. Ed said they felt someone from the boat segment of the business would be better qualified. He then proceeded to tell me that the reason he came down was not for a plant tour, but a ruse for talking to me about being a candidate for the purchasing slot on the PWC team. I was totally pumped. It was exciting to be recruited. A colleague of mine had suggested to him that I might be a good fit.

MARINE POWER PRODUCT GROUP PURCHASING

So, I went through the proverbial interviews, back and forth flights, visits to show my family the area, and so on. I got the job. I got a raise. Considering the high taxes in Illinois compared to Florida, my raise was a wash within a couple of dollars. I was now a technical engine components buyer. Cool. I was pumped. This was going to be quite a cultural change from boat building. There was a dress code mandating a sport coats and tie. This was very different from the boat side of the business where the uniform of the day was a polo shirt with company logo and jeans. Having earned an MBA earlier in my life, I was all over this. Hey, what the heck, life is short. Now I only wear suits when I have to. So in January of 1994, I moved my family to Kenosha, Wisconsin, from Sarasota, Florida. We went from 75 degrees and sunshine to 20 degrees and a foot of snow.

OMC was the manufacturer of Johnson® and Evinrude® boat motors. It was a traditional Midwestern manufacturing company. As I went around the corner, I could see the building where marine power products group was located. It was just feet from the shore of Lake Michigan. The wind was blowing the snow sideways and it was cold. I stepped out of my truck in my new suit and headed into the building. There was a lobby with a telephone on a table with telephone book next to it. A few suppliers' representatives were in the lobby. I looked up Ed's name and called him. He came out and welcomed me. He quickly introduced me to my senior buyer, Bob. Bob was responsible for bearings, cables, carburetors, and castings.

Bob was a likeable guy. Almost everyone liked Bob. He was southern, played bluegrass music, and was hyper. Every summer he had cookouts at his home in the country, and they were quite the ticket. Lots of beer, music, great food, and great fun. Bob was a hard-working senior buyer. He handled his accounts with discipline. He processed his requisitions daily, maintained blanket orders, solved supplier issues, and dealt with any plant issues quickly. Actually, our work styles were similar. Bob had built really good relationships with the expeditors in the plants. During a future trip, I noticed that someone had reserved a parking spot for him with a little white sign in front of one of the plants. That told me a lot. But hey, manufacturing is about people and processes. I took notes from Bob. He taught me a lot.

I remember on one occasion we were discussing pricing with representatives from Torrington Company bearing in our conference room. The reps presented us with a list of pricing and indicated that they felt good about being able to present the plan. Then all of a sudden, Bob tells them where they can put the pricing, and he did not hold back his words. He cussed them out and stormed out of the room. I was taken aback. I apologized to the Torrington reps and asked them to wait while I found out what was wrong with Bob. This was totally out of character for him, and I wanted to talk with him to find out what was going on.

I finally found Bob outside smoking a cigarette. I asked him if he was all right, and he laughed at me. "Why wouldn't I be?" he asked. I told him that I had never seen him like that. He said that was because I had never watched him negotiate. I could not believe it. He told me to go back inside and tell them that he would be back in when he calmed down; and that if they were smart, they would wait. So a while later, Bob reentered the conference room, apologized, and advised them that if they wanted to continue our business, they had better come up with some better pricing. He sent them off, and a couple of weeks later they presented a much better pricing quote. What an adventure!

Purchasing was in the building that was the original Johnson Outboard® manufacturing plant. I immediately discovered the area of the building that housed a manufacturing area that fabricated service parts such as pistons and some other castings. It was a good-sized factory, and I snuck away from purchasing nearly every day to explore. It was an absolute adventure. There were several aisles of machining centers, and the smell of machining oil was in the air. There was a great sense of pride and professionalism displayed. Along the aisles next to the machines, there were uniformly placed baskets of die-cast aluminum parts in different stages of manufacture. Each basket had a printed traveler with routings, etc. OMC had a high level of shop-floor discipline. It was impressive. However, with all that said, that is not to say that they were Lean. OMC factories were impressive, full of highly skilled technicians, clean, organized, well maintained, had a strong emphasis on quality—but they were not Lean.

In the back of the building was engineering. This area had cell after cell of engineering technicians assembling or disassembling engines to test components or evaluate tested components. There were load cells, workbenches, and engines everywhere, with test barges outside the door in the harbor. At all times, engines were affixed to the test barges and ran continuously. This was done to conduct reliability testing for the engines themselves and to test components. The fuel/oil mixtures were manipulated to simulate a longer testing period. Generally, supplier components would have to successfully perform on some of these engines for sometimes more than a year to be eligible for use on an engine. Reliability and durability testing or prediction can be tricky. The quality department was using "Weibull" distribution analysis to make predictions relative to product life, time to failure, etc. It was fascinating. There was not a day that went by during my time at OMC that I did not learn something valuable about manufacturing. It was a truly rewarding experience.

In the middle of the building coming through the ceiling were the legs of a water tower that rose above the building. My desk was only a hundred feet or so from the legs coming in through the roof. On the water tower were the words "Johnson Outboards" with the seahorse logo. I got such a kick out of working in the factory that Ole Johnson himself had walked the floors of. Even more fascinating to me was a very large room that must have had 200-plus Johnson and Evinrude boat motors in it. These engines were accumulated over years and were kept for research and reference. It was basically a private museum. And what a collection it was! I, in particular, appreciated these engines because as a kid I had assembled more than one outboard motor from broken-down motors put out for the junkman. What a blast it was to walk through this room and survey the outboards.

My work at OMC was literally nothing but fun. The drive into work from Kenosha to Waukegan provided a small tour of the industrial Midwest. The silhouette of a large and old factory stood in the distance with big scrubber stacks standing proud against the silvery-gray dull sky over Lake Michigan. What an awesome sight it was, yet a subtle reminder of what used to be. As I turned onto Sea Horse Drive, I looked up at the Johnson water tower. This was an icon and fun to see. As I continued on, I went by Outboard Marine Corporation's corporate headquarters. A flashback of the OMC logo and "Made in Waukegan, Illinois" words that used to be on the boxes of the OMC stern drive engines I used to install at Sea Ray went through my mind. I was taken aback by how small the corporate building was. Nothing like the "Taj Mahals" you often see these days. Maybe it looked small because it sat right in front of the OMC Die Cast foundry. At one time it was the largest die-cast facility in the United States. There were blast furnaces, dies, and castings everywhere, transfer lines, the smell of machine oil, and people driving around on golf carts. And as always, it was a very clean and organized facility.

GETTING TO KNOW THE GEMBA: THE PRODUCT AND THE PEOPLE

As a boat builder, I installed components on small cruisers, worked with gelcoat, fiberglass and resin, molds, and a variety of hardware; so I had a fairly strong familiarity with the product and components. However, as I transitioned to larger boats and yachts, all the way up to mega-yachts, there were different components involved. In every role I had, whether as a buyer, materials supervisor, plant manager, or vice president, any time I was handed or involved in a component or material I did not recognize, I went right to the factory floor and found out what it was and how it was used. This also gave me an opportunity to meet the people on the factory floor, the people who are the keepers of the most important knowledge in the company. Building relationships with operators is not only fun, but also they are a huge resource in problem solving and product information.

During my first week on the job at OMC, I also noticed the service training school in the building. It was called OMC University. They brought in outboard service technicians from all over the world and trained them on how to service, troubleshoot, and repair outboard motors. As a kid, I had several outboard motors and had spent a fair share of time working on them. Even with that familiarity, I was already hearing the engine lingo in the company and I had some learning to do. I asked my boss if I could sign up for that school. He said, "You're a buyer!" I told him that I could not possibly go wrong knowing our product, and the school lasted only two weeks. He allowed me to go, and I actually think he liked that. So two weeks later, I was a certified outboard technician with certificate in hand. Getting to know the company's product was essential. It allowed me to become more valuable and contribute at a much higher level.

OMC–Burnsville

Once I completed the training, I set out to understand the components that I would be purchasing, the plants these components were used in, and where on the engines they were used. The majority of my components were used at OMC– Burnsville, OMC - Andrews, and OMC - Calhoun plants. My boss had mentioned that the corporate jet flew shuttles from plant to plant twice per week. I asked him I if could visit the plants to which I supplied components. He said, "Sure." So I made arrangements to spend two days in each of these plants. I went to each plant and introduced myself to the plant manager, the materials manager, the Expeditors, and anyone else I could meet. I spent the majority of my time on the factory floor observing how and where the components I purchased were installed on the engine. I remember a plant manager asking me, "Who are you again, and what are you doing here?" I said, "I'm Bill, I'm a buyer, and I will be sending components into your plant so I thought I would try to get to know you, your people, and how my parts fit into your operation. It will help me do a better job." He looked at me and shook his head, but I could tell he was impressed.

Linda Thomas was the lead expeditor at OMC–Burnsville. She was an awesome professional. She was one of those women who were a key contributor in the factory, but you also just knew that she managed her home with the same urgency. A good person all around. You talk about energy. Her job was to ensure that the flow of component parts into the plant met the production demand. This factory assembled power heads and was the life-blood of the Calhoun assembly plant. They received the blank power heads from the lost foam plant, machined them, pressed in the cast-iron cylinder sleeves, and installed a myriad of other parts and sent them on to Calhoun. If any one of those parts was short stocked, production stopped, and it was Linda's job to make sure that did not happen. She was great to work with, but when you told her something, it had better be right. It was the Monday after that visit that I started to call her at the start of each day.

This factory was another adventure. It had machining centers, transfer lines, and sub-assembly centers. It was intense. The amounts of equipment, machinery, and technology were amazing. Everywhere you went, there was cutting oil or fluid squirting on something being cut, or chunks of engines here and there with operators installing more components, tightening screws, boxing things up. There were literally thousands of tasks and actions occurring simultaneously and in harmony. Everything had a tolerance, and quality was key. Nearly all components were sample tested and measured on a coordinate measuring machine. This was serious manufacturing. At the end of the plant, the power heads were affixed to a specialized pallet and shipped to OMC – Burnsville to be married up to lower units and other units to make it an engine.

OMC – Calhoun

My fascination in being part of an engine manufacturing company continued. I was just taking in everything. It was an absolute blast! OMC – Calhoun was the final assembly plant for Johnson and Evinrude outboards. Components came in through a row of loading docks in receiving. Thousands of components arrived each day. Just inward from the docks were a series of expeditors. These folks maintained awareness of stock levels of engine components and stayed in constant communication with suppliers either pushing scheduled component deliveries in or out with the flow of production.

After unloading, components were either staged in areas or delivered to point of use in the plant. Overhead in the factory was a conveyor transporting lower units, power heads, engines, and cowlings from place to place. This included the painting area, which was quite an operation. Outboard engines live in saltwater and other harsh environments. Avoiding corrosion required a several-step and specific painting process. So, you would see components on the overhead conveyor going from area to area in different stages of the process.

Once the engines were assembled, they went into testing cells where they were started up and ran to specific quality testing to ensure operability. This was quite an operation. It was a large structure with grating floors, cells for engines, and hoisting and moving apparatus. After testing, the engines were cleaned, prepped for shipping, and put into boxes. They were stored in a warehouse and shipped out to fill customer orders.

As usual, I floated around everywhere. I made my way into the daily production review meetings. These meetings were just slightly this side of tense—not terrible, but it was not a jovial atmosphere either. Like the Burnsville plant, there was a lot going on in this plant, and keeping production going and solving problems is not easy. It just so happened that during the week I was there, the senior VP of manufacturing was visiting the plant. His name was Sam and Sam was *the* man. He had responsibility for all of the company's factories and operations, and that was a lot. One day he happened to attend one of the production review meetings I was in. As the meeting was going on, each person gave the status of their area, what issues they were dealing with, etc. They were bringing up quite a lot of issues too, I noticed. I remember thinking to myself that these guys were so much more professional than the management at boat plants. Raised voices and profanity were regular occurrences where I had come from. This was obviously a whole new game. A few minutes later, someone asked Sam if he had anything. Being the big guy, all ears were on him. So he said, "You know, I have listened to you guys go around the table and I have to tell you, all I hear is a bunch of #$@!ing bullshit! Nothing but a bunch of excuses, and I suggest you all pull your heads out of your asses and get something done! We get paid to make engines, not excuses!" And he got up and walked out. OK, well, I guess there were more similarities than I had thought! True story. Some four hours later, Sam was terminated. It was his time. I heard a lot of good things about Sam over the years. He was very well respected, and you know that to rise to the level of managing nine high-volume, integrated factories, he had to have something going for him.

After returning to the office, I was visiting with some of my new teammates. The subject of me going to the plants came up. They acted surprised. I asked how many times they had been to the plants. Most of them had never been to the plants at all. And some of these guys had worked there for more than fifteen years. None had ever gone to the service school either. I thought that was disappointing. These were good guys, and they were missing out. I could not imagine supporting factories and never visiting them or meeting the people who worked there. Within six months, I had ten times the exposure to the plants, processes, and people they supported than they had gained in ten to fifteen years with the company.

Coming up from the factory floor during the start of my career, I understood how important having parts was. Factories move fast, so you have to stay on top of it. Each morning, I would come in at about 6:00 a.m., grab a cup of coffee, and call the expeditors at the three plants I supported to ask how things were going. If they needed something, I knew it. I noticed that there were only a couple of other buyers who did this out of a total of about twenty. I would see them go to work, and if a plant called, they responded; but it never worked as well as a proactive approach. I have always made it a practice to go to my internal customers and not make them come to me. I feel this is part of the "Lean Culture." A great sense of urgency, and doing everything that maximizes value to the customer.

Lessons Learned

- The value of understanding your internal customers, the Gemba you support, and the components you supply cannot be overstated.
- The value of going out on the floor of the factory or into the business is great.
- Proactive always trumps reactive.
- If there is manufacturing, there is passion.

JET SKIS

As a member of the personal water craft (PWC) team, I attended all team meetings. The engineers were so impressive at OMC. They were young and very smart. I was amazed that these guys designed engines and all the gadgets that make them up. The engineering team had assembled a rough prototype called the "Shock Wave" and it was great. One day in February, one of the guys on the team donned a wetsuit and took it for a spin in the harbor…OK? We were all standing outside on the docks with our teeth chattering, and this guy was spinning around the harbor in water right out of Lake Michigan in February. There was ice on the shores and on the PWC. Seeing him kick up water and spin around in a machine that the team had built got everyone pretty fired up.

A few weeks later it was decided that we would be going to Stuart, Florida, to the OMC test facility to do some testing. The whole team would be going, including me. The trip was to be for a week and I packed accordingly. They had scheduled flights on the corporate jet for us. The jets were Hawkers. They were short, stout jets with seven seats and took off like a rocket. We all showed up at 7:00 a.m. at the Waukegan airport, stowed our luggage, and away we went to Florida. This was such a blast! We also had great conversation about the project on the way down to Florida.

We checked into a hotel and then made our way to the test center. The uniform of the day was swim trunks and flip-flops because we would be testing. We showed up at the test center and walked through toward the water. On the way in, there was a nice tribute to Ralph Evinrude, as it was the Ralph Evinrude Test Center. Out back at the docks there must have been about ten PWCs, jet skis, etc. There were only about eight of us, so that meant that there were plenty of watercraft. Our mission was to continually test drive the jet skis, and each time we returned, to write down what we liked and what we did not like about each one. There was a table full of fruit and snacks, along with coolers full of drinks.

Twenty minutes later, I was riding a Yamaha Wave Runner in the Intracoastal Waterway alongside my teammates, each riding a different watercraft such as Sea-Doo®, Kawasaki®, and so on. I was getting paid to ride jet skis! I could not have had more fun. Mega-yachts were passing by, boaters, as well as other jet-skiers. This was unbelievable! We did this for three days, and each night we went out to a great restaurant. The Yamaha® craft were fast, almost uncomfortably fast, but awesome. The Kawasakis took off with almost a "wheelie." That was neat, but they were not as fast as the Yamahas. The Sea-Doos were versatile. They took off quickly, and rode and handled nicely. They were all fantastic machines. We made notes of our likes, dislikes, and ideas, and they were recorded for review upon return. To this day, these were some of the most fun days in my entire career.

We boarded the jet, tired and sunburned from three days of "testing." As usual, everyone was telling stories and "shooting the breeze" on the jet while flying back home. Then we flew through some thunderstorms. The jet dropped a couple of hundred feet about ten times; lightning bolts were going off outside the windows. There were beepers going off left and right in the cockpit, and the pilots were bouncing up and down like we were. You did not hear a peep out of anyone until we hit blue skies. It was the craziest ride I have ever had on a plane or jet to this day. It was nice to land.

A while later, the company decided to scrap the program due to market conditions. Luckily, I had taken on every account that anyone did not like and had a pretty full portfolio of commodity items to create a job (that proactive thing again). I purchased bearings, control cables, carburetors, valves, and a host of other engine components.

One day I was called into the purchasing director's office. He said he had a project for me. I was not sure what it was. It turned out that during the PWC project, OMC had outsourced some components to a supplier in California. They were in possession of some valuable tooling used to make components for more than just the PWC effort. The supplier was in bankruptcy, and our job was to secure the tooling. I was asked to join an associate from engineering in going to the supplier and getting the tooling.

We showed up at the supplier. They were a small-volume PWC manufacturer. Actually, it was a well-laid-out plant. 5S was evident. It was very organized, and it was set up for a very efficient flow of product through the process. They were using pre-formed fiberglass and a resin transfer system that was exceptional. There was a trustee involved, and he and the owner were going round and round. We used some real diplomacy to get the tooling. And we succeeded. However, we had scheduled three days out there. It was early in the afternoon and our flights were not until the next afternoon, so we had some "time to kill." We headed to Huntington Beach, had dinner, and checked out the surfers.

My associate asked me if I had any ideas on what to do. Well, the OJ trial was either in process or just finishing. I suggested we go to Brentwood and check it all out. We went by the Mezzaluna restaurant where Ron Goldman and Nicole Simpson had eaten just before the murder. I got a map. We drove to the condo on Bundy where the murders had occurred. We then headed for OJ's home on Rockingham. We drove straight there and made it in ten minutes. There was fencing put up around the home, and I drove right up to the telephone at the gate that Mark Furman had been pictured talking on in the newspaper. OK, that was fun—weird, but fun. Back to Waukegan.

Lesson Learned

- It takes more than just a Lean factory to be successful.

CYLINDER SLEEVES

Bob M. called me to his cubicle one afternoon and advised me that I would be traveling to a supplier. Evidently, OMC and its chief competitor Mercury were buying piston sleeves from the same small supplier in Muscoda, Wisconsin. Mercury had a full-time purchasing person in there, and so did OMC. Each representative was there to make sure their company got its fair share of the capacity. Obviously, the number of engines you can produce is in some way a function of the number of piston sleeves you have on hand.

After driving through the back roads of Wisconsin for most of the day, I showed up at the factory. There was minimal parking space and I had to park on a patch of grass. It was cold and rainy, and I walked through mud on the way to the entrance. The factory was owned by a very interesting gentleman. He was a Bosnian Serb immigrant and a very passionate man with a temper. His broken English, passion for his country, and some alcoholic beverages over dinner made for some interesting conversation. It was fascinating to hear him cite the history of his homeland and talk about who was fighting whom and for what. I did have a fleeting thought of appreciation for living in America away from that type of fighting and disruption of life. We finished dinner and headed to hotels for the night.

It was a relatively small, plain-looking factory. I came into the lobby and there was a lady behind a window who turned toward me when I approached. She pointed me toward the conference room. The people did act a little nervous. It was not because of me. I set up in the conference room with my laptop and introduced myself to my counterpart. He was a good guy. As usual, I got right out into the factory. It was dark, very dark, only having about a third of the lights that it needed. You could smell machine oil, and there were saws for cutting iron, Takasawa® machine centers, Bridgeport® machines, scrap bins, and baskets of pistons in different stages everywhere. Some of the equipment was old. It was not the cleanest factory. There was debris and items, etc., spread in the fringes of the factory. As I came around the corner, I encountered one of the employees. He approached me and asked me how I was doing. I said fine and we made some small talk. He was a tall man wearing suspendered coveralls with a "chew-in." He was also missing some teeth, actually most of them. I mentioned to him that I had heard they were having some quality challenges. He chuckled, and I will never forget this, and said, "Huh huh, our biggest customer is the scrapyard!" Oh boy. It was unbelievable that a small, loosely managed factory in the middle of nowhere was supplying critical components to state-of-the-art factories of world-class engine manufacturers.

Being the passionate manufacturing professional that I was, I was pursuing APICS certification during this time. One of the six modules in APICS is Just in Time. In order to earn the CPIM designation (certified in production and inventory management), one had to pass five multiple-choice examinations. They were tough, and in-depth preparation was not an option. I had taken a three-day Just in Time exam preparation course in Chicago and was reading *Just In Time for America* by Henry Wantuck, so I was all pumped up. During my training and preparation, I encountered the topics of process flow and factory layouts, etc. Because I was at the factory full-time one week on and one week off, I had plenty of time on my hands each day. So I decided to create a flowchart or mapping diagram of how materials flowed through the process. It was a rudimentary Value Stream Map. I taped four pieces of grid paper together and assigned so many feet per square for scale. I drew the equipment on the map, recorded the cycle times of each piece of equipment, and noted the quality issues at each stage. It was my first one. It needed some work…but not bad for a first time. The process for making cast-iron piston sleeves started with cutting lengths of cast-iron tubing with an electric saw, followed by broaching and milling, and then using a Bridgeport machine to cut finger ports in the sides of the sleeves. These finger ports were a feature that OMC was proud of, although they were hard to create. Getting the finger ports in properly and maintaining the roundness of the sleeve were the two critical-to qualities.

Once I mapped out the process, I took a look at product quality issues. There was a high scrap rate. One thing that stood out was that they only measured quality at the end of the process. I talked with the quality manager about the product specs. They were looking for what was called a "tri-lobal" condition—or better yet, the absence of it. The sleeves had to be round, very round for the piston to slide through them effectively. Just as important was that there could be no gaps between the sleeve and the powerhead into which it was pressed. A gap would interfere with heat transfer and cause a hot-spot in the cylinder wall.

I twisted some arms and convinced them to measure quality at each step in the process to stop defects from passing from one operation to another. They were experiencing a great deal of difficulty cutting in the finger ports, which were basically an oblong hole in the side of the sleeve shaped like the tip of a long fingernail on each end. Basically, the process was not capable and/or had a high defect rate. Too many of the defective units were making it to the plant, so that was one of the areas we were monitoring while there.

We did apply some basic root cause analysis and found that the cutting edges on the machine needed to be checked for alignment more often and adjusted properly. It helped, but it did not improve to an acceptable level. The process just was not capable. Ultimately, OMC decided to develop an in-house capability and a year later moved them into the Burnsville, North Carolina, plant.

Lessons Learned

- It does not do you any good to have great internal quality processes if your suppliers do not have them also.
- Passing on quality defects in the process always makes things worse.

MORE STANDARD WORK

Settling in to the Marine Power Products Group (MPPG) purchasing department was a challenge. Not unlike any other job, the usual having to learn the new computer software, the people, the way things are done, procedures, all that. Again, this was an old manufacturing culture, and you paid your dues. There was a pecking order...and lots of politics. I was taking it all in. On the other hand, my senior buyer was great, but he was also busy. He showed me how to process requisitions, how to request a quote, how to process a price change, how to create a blanket order, all the things that buyers needed to know. OMC had a system called MCS, for manufacturing control system. It was an AS400-type system. It was robust and did a great job, but it was slow on entering data, etc. Heaven help you if you accidentally hit a key twice. It would beep in a tone that would put a dog on its knees, and then you had to go through the seventeen-step log-on process (sarcasm, but not far from it) and once again, heaven help you if you messed up.

I am a slow learner; but once I know something, I know it. I did not want to bog down my boss by asking him repeatedly to show me how to do things, so I would ask others. I remember asking one of the more senior buyers if he could help show me how to process some transaction. There was quite a pecking order there, and you could tell that he thought he was pretty high up on it. So he said, "OK, just hit this key, enter this, hit this key, go to this screen, and then do that." Then he turned and walked away. Oh boy! After going through this a couple of more times with different people, I started asking some of the younger buyers to show me things. When they did, I started printing screenprints and highlighting fields that needed to be filled out for different transactions and making notes to help me the next time I did this. A lot of people were very helpful, but there were a few who gave me the "hundred-miles-an-hour lesson" and walked away. It was interesting.

After a while, I had quite the binder of work instructions on my desk. I used document protectors and set it up quite nicely. The next thing you know, people were stopping by my cubicle and saying, "Hey, don't you have some book on how to do things?" I would say, "Sure," and let them borrow it. I would sometimes ask, "You don't know how to do that?" They would say, "Not really; I always guessed at that. I would like to see what your book says." Then we got a new vice president of supply management. What do you think one of the first things he did was? "Hey, do you have some book on procedures?" Even some of the guys who gave the "hundred-miles-an-hour classes" started asking.

In addition to an inconsistent knowledge of processing system transactions, no two buyers did their jobs the same way. When I would ask how something was done, someone might say, "I do it this way but Denny does it that way and Ray does it another way." Every buyer had his own personal process. So, the bottom line was that a team of people purchasing over a half a billion dollars per year of raw materials and engine components were guessing how to do things! It was crazy to think that the team supplying the components to produce over a thousand engines per day was guessing at any part of their job. This is not a knock because they were very experienced buyers.

Standard work practices were one of the key focus points of the new leadership, and it was needed. You cannot control a process that is not defined, or followed, consistently.

Lessons Learned

- Standard work and/or procedures are invaluable in training new people.
- In the absence of standard work or procedures, a huge degree of variability can be happening just under the surface of the process.
- In the absence of standard work or defined processes, each operator will most likely develop his or her own process, which creates variability.

8

Japan

Outboard Marine Corporation (OMC) was an old, established company. Like many other American companies, how long you had been with the company was a big deal. People literally wore it on their sleeves. Actually, they wore it on their ties and wrists. Employees received a five-year tiepin; at twenty years, you got a watch; and you may have received a new tiepin at ten and fifteen years. They wore them like sergeants wore hashmarks on their uniforms in the military. When people introduced people, they told how long they had been there: "This is Jim, he's been here seventeen years, and Bob, he's been here twenty-two years. How long have you been here?" In the Lean culture, pride is taken in the number of jobs or skills an operator has. They are held in high regard. One of the problems in a lot of American companies is the emphasis on how long you have been there, as opposed to the number of skills you have and ways you contribute to the company. Think about it: we do not care how long our doctor has practiced. We do care about his or her skill level and success rate. Experience is nice; however, skill and performance constitute the real value. One of the tools of Lean is a skills chart on which employees are listed on one side and skills along the top. An "X" is placed in the appropriate box for each skill an employee learns and is willing to apply. This is a great tool for moving the focus from seniority to skill and peformance.

During this time, the OMC quality department was implementing a supplier quality system survey process. If an engine component on a drawing had a diamond symbol somewhere on it, that meant that that condition or attribute was a critical dimension. Any supplier supplying a component with a critical characteristic had to submit a process capability study verifying that their process was capable, establish a control plan including a sampling plan, and pass a supplier quality system survey conducted by one of our quality engineers.

At the time, I was purchasing reed valves from Indonesia, Taiwan, and Japan. I started thinking that it would be a blast to travel there and conduct quality surveys. While I was doing that, I could study and observe Japanese manufacturing techniques first-hand (all three locations were Japanese owned). So I took it upon myself to participate in some quality audits and earn the ASQ (American Society for Quality) certified quality auditor (CQA) certification. One day, I approached my boss and asked him if I could travel to my suppliers in Taiwan and Japan to do quality surveys. He quickly pointed out that I was not qualified. I then showed him my ASQ CQA certificate. He just shook his head.

There were a couple of other situations occurring with regard to our supplier (NOK Freudenberg). In addition to needing a supplier quality system survey, OMC had become concerned about a rumor that they had built a new state-of-the-art factory in Taiwan where they were making similar product for Mercury while manufacturing ours in the older factory. As I look back, I realize how inconsequential this was—the quality of a product really is not a function of the factory it is produced in. Given an acceptable design, it is a function of the capability and control of the processes the product is manufactured in. We were also experiencing some quality defects on some of the reed valves we were receiving from them. And finally, we had a proposed design improvement we wanted to talk to them about to improve the product.

Reed valves are an integral and key component of a two-stroke engine (see Figure 8.1). They function as a mechanism to control the flow of the fuel/air/oil mixture between the carburetors through the crankcase into the pistons via ports in the side of the cylinder walls. Reed valves are made from machined aluminum die-cast parts with stainless steel reed valves

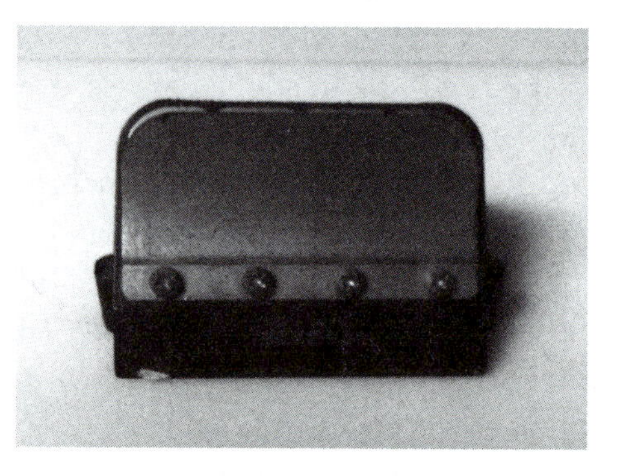

FIGURE 8.1
Two-stroke engine reed valve.

fastened to it covering passages for the fuel/air mixture to flow through. There is one reed valve assembly for each cylinder on the engine, and they are mounted on the power head behind the carburetors.

The critical characteristic of the reed valve was the flatness specification of the face. Any leaks—even of the smallest nature—would reduce the horsepower and performance of the engine. The valves, which are quite a mystery in and of themselves, must seal tightly when closed and flow like crazy when open. Any leaks or lack of flow when needed will greatly impact the performance of the engine.

A higher than acceptable number of units from recent shipments were failing incoming inspection tests. Samples from each shipment were given a flow test and a vacuum test. The units were primarily failing the vacuum tests. The company now had to test 100 percent of the units in order to keep production going. Resolving this quality issue was on our agenda during the quality system survey trip.

A few weeks later, I was on a 747 headed for Tokyo. We got there for the weekend. We checked into the Tokyo Prince Hotel and proceeded to explore. We went by the Russian Embassy. It was really a site. It looked spooky, like something you would see in a James Bond movie. It just happened to be St. Patrick's Day, and we stumbled upon the Hard Rock Café in Tokyo and got to drink green beer. As we walked along the streets toward the Hard Rock, we passed several "Pachenko" parlors. Pachenko parlors are full of bright—and I mean bright—lights and look like casinos. They are lined with people staring at the lights and playing the game.

I was totally pumped about the opportunity to go to Japan and see their Lean manufacturing. Let me tell you just how excited. This was during the time that China was firing missiles across the Taiwan Strait. There was also a big election coming up. Someone asked me if I should not postpone my trip. I did not. Given the same situation today, I might act differently.

On Monday we headed up the mountains to our first factory. We drove through tea fields with Mount Fuji in the distance. It was great. We arrived at the factory and were greeted with the utmost respect and hospitality that the Japanese are known for. It turned out that this was not the factory in which our product was produced. Some senior management was located there, and they wanted to meet and greet us. I have so much respect for the Japanese people and their attitude toward manufacturing. They are so respectful and passionate about quality. At least that was the case with the team at NOK Freudenberg. So we met, bowed, shook hands, small-talked, and finished up.

We then moved on to some smaller factories making some of the components. This is where I first began to realize some of the simplicities of Japanese manufacturing techniques. We pulled up to a building with corrugated galvanized steel on the outside. It was draped in some sort of ivy-like vine and plants. I was not sure what the building was. One of the representatives opened a sliding door to reveal a small factory. You could see several assembly stations all laid out the same way. Kanban cards were located throughout the small factory. 5S was obviously in place. There was a place for everything, and everything was in its place. Standard work and certain specification information were present at each workstation. What surprised me was that I did not see a personal computer anywhere in sight. The equipment in use was older, yet seemed to be working. On the benches was some of our product. However, the primary components we were concerned with were not there. They would be in the Taiwan factory. This little factory was very impressive. We audited it and prepared to move on.

On the way back to Tokyo, the company representatives were excited about stopping in some small, out-of-the-way village for us to have green tea with some folks. They had prearranged the stop and were really excited to show us this part of their culture. I actually could not have been more impressed. I am not even sure who these people in the village were, but they brought out some special tea leaves, elegant teacups, and served us green tea. I believe it was during the first tea of the year. So we bowed, drank tea, and conversed about this and that for half an hour in a small village in sight of beautiful tea fields.

That night, we went out for dinner in the Rappongi district of Tokyo. We went to a German restaurant. It was great fun. Over each table was a big bronze bell about two feet in the air. The deal was that if anyone bumped the bell and made it ring, they had to buy a round of beer for the house. The bell seemed to be in a place above the table where there was little risk. Not! Sure enough, after a couple of beers during conversation, up goes a mug and ding goes the bell. I did think it was strange going to a German restaurant in Tokyo.

Mr. Yamata

Late Tuesday night, we arrived in Kaoschung, Taiwan. There was a huge upcoming national election, and there were banners and flags as far as the eye could see. We checked into the hotel and proceeded to go out to dinner. We sat at a table with a top that was basically a carousel full of food with a ledge around it for plates. The idea was that when something went by that you liked, you were supposed to grab some. It spun many times before I grabbed something. I love ethnic food, but the smell and looks of that food did not work for me. Luckily, breakfast in the hotel was better.

The next morning, our supplier representative and interpreter picked us up and transported us to the Taiwan factory. As soon as we pulled up, we noticed the factory was recently built. Our fears of this being a substandard operation were immediately reduced. Inside we met the plant manager, Mr. Yamata, and his staff (see Figure 8.2). Again, like disciplined soldiers in an IG (Inspector General's) inspection, they located and presented any and all data we requested as part of the audit. It was like a sport to them, and my respect for them grew by the minute. It was pure manufacturing, and it was great. Plant manager Yamata was highly revered and respected, and quite a serious guy. (I am thinking Taiichi Ohno 9) I will say that he looked at us with a minor glare now and then. I noticed that he was not taking notes to learn how we Americans did it. More on that later.

FIGURE 8.2
Our group involved in a quality system audit. The author (standing); Tony, his teammate, kneeling directly in front of him; and Mr. Yamata with his hands on his hips.

FIGURE 8.3
The author on the left and Tony, his teammate on the right, outside the factory we were auditing.

The factory, as the previous one, was clean, organized, and well laid out. However, everything was newer. I did notice that the equipment was all relatively small compared to that in most U.S. factories. That stood out to me. It was like miniature versions of normally large equipment. Small conveyors, small rinsing baths, and small plating operations. We thoroughly reviewed their quality system, and it was impressive. What was just as impressive was their attitude toward quality. It was expected in everything. They used simple cards, binders, folders, Kanban, etc. to control nearly everything. Nothing high-tech, all very simple. The technology they had was in the equipment they used to manufacture their product, but not to track and control shop-floor activities. We were so impressed we had our picture taken outside the factory (see Figure 8.3).

During their breaks, the employees slept. They lay down anywhere or put their heads down on their equipment. Management went to great lengths not to disturb them. They reminded us several times: "Shhhh.. We don't want to disturb our employees on their breaks." We literally tip-toed through the factories at break and did not talk. This was a pretty big allocation of respect for their employees in my eyes.

Next, we were scheduled to meet with plant manager Yamata and certain members of his staff to review the quality issues and discuss the design modifications. The quality issue was that a flatness specification for some of the reed valves was being missed too often. The reed valve is fastened to the engine power head, and it must be highly flush with it to avoid leaks, which would cause a significant power loss. As we began to discuss the issue with them, a couple of things became apparent: first, how well they worked as a team to solve a problem, and second, that they used a problem-solving process. One of the quality staff members immediately sketched out an Ishikawa or fishbone cause-and-effect diagram. It was as natural as breathing to him. He did it with the ease with which a young kid in America would create a tic-tac-toe matrix. He wrote down man, method, machine, and materials. There was quite a bit of discussion, and then we visited the Gemba where the valves were manufactured. We traced every step in the process. We did man, method, machine, materials, and PDCA (plan, do, check, act).

To achieve the flatness on the bottom of the components, about a dozen at a time were dropped into a fixture something like eggs into a carton. The reed valves were smaller than your fist. They were clamped down and held in place. A large grinding wheel spun over them to grind them flat. After much discussion with operators, team leaders, and us, it was determined that the root cause was that the components were moving around in the fixture and this yielded uneven surfaces. The fixture was modified with a clamping device to hold the valves in place during grinding and the problem was solved. Plant manager Yamata expressed pride as his team solved the issue. It was pretty cool.

Another concern was put to rest as we reviewed the manufacturing process, and that was flow and vacuum testing. NOK had developed a really capable and awesome machine to test the vacuum and the flow. The machine was capable and calibrated, and each valve was placed over a hole with a vacuum draw on it, and then clamped down and a flow put through it. This was a welcome sight and finding.

Are You Sure He Is Not Mad at Me?

OK, it was time to discuss modifying the design to an improved state. This supplier had produced the parts to spec and in the design we had approved. They also manufactured the tooling and fixtures. The interpreter initiated the discussion and as he talked, plant manager Yamata's face became stern and his eyes were glaring. When the interpreter finished speaking, Mr. Yamata began to speak in Japanese, which I could not understand. But I was watching his body language, and he actually looked angry. He looked like one of the pilots in the movie *Tora! Tora! Tora!* I asked the interpreter if Mr. Yamata was angry. He said, "Oh no, he is just concerned about how to make the changes."

In a loud voice, Mr. Yamata said something like, *"Hun dun huncheechun, huncheehunhunbao! Ah un bun ho hao chi si bon bun bao! Hus hen chao dun chin hunhunbao! Bao, Bao! Hun dun hun!"* (There is absolutely no disrespect meant in describing this situation, as I so much admire these people.) And he said it with a mean look and glaring right into my eyes. This guy looked very angry! I again asked the interpreter, "Are you sure he isn't angry with me?" "Oh no," he said. "He's just discussing the change." I asked, "Well, did he agree to do it?" The interpreter said, "Hold on," and talked to the plant manager some more. Same thing; looking right at me, Mr. Yamata said something like, *"Huncheechun, huncheehunhunbao! Ah un bun ho hao chi si bon bun bao! Hus hen chao dun chin hunhunbao! Bao, Bao! Hun dun hun!"* I said to the interpreter, "I am telling you, he is mad at me!"

Looking back, I know that the plant manager was saying something like, "A typical American company. They ask us for a design and we give them a design prototype with the opportunity to change anything. But no, they approve it completely. We buy or make tooling and make jigs and fixtures, and now they scratch their heads and decide they want to improve the design. If we make new tooling, we will have to raise the price and they will not accept that. Why do Americans try to achieve quality *after* the design and not *during*?" And we all knew he was right.

At the end of the day, the supplier refused. The interpreter never admitted that the plant manager was mad at me. It is strange, but I would like to have a beer with plant manager Yamata someday. I think it would be interesting to hear his honest perspective. The election and weather caused flight delays, and we had to cancel the trip to Indonesia. We would have to go another time.

Closing Out the Audit

We continued the supplier quality system survey. We checked capability studies, gage reliability and repeatability, gage calibration, separation of defective material, materials certification sheets, equipment maintenance. Honestly, we drilled them very thoroughly. Once again, I was totally impressed with the attitude of the team leaders. Each time we asked for something, such as an equipment record, etc., they would say, "Ah yes," and take us right to it. Usually some sort of simple card from a file with minimal but required information. They would present each document with a great sense of pride and accommodation. They were proud and took their jobs very seriously. Actually, sometimes I thought they should "lighten up" a little bit. But after work, they are just as light as they are tight during the day. They took us to the karaoke bar and it was all fun.

They passed the survey with great results and we took some pictures, and I feel not only parted as professional associates, but as friends, as fellow soldiers on the "Battlefield of manufacturing." To this day, that trip is one of my favorite experiences in my twenty-five-plus years of manufacturing.

A few weeks later, we received a call from our plant informing us that the Indonesian reed valve supplier had shipped a batch of valves that failed incoming inspection. There were quality issues and, more specifically, the vacuum test. The valves were not sealing properly when closed. I regretted having to cancel the trip, although I was a little nervous about going to Jakarta.

Mouth Test

I called the supplier in Jakarta. I had to call at some odd hour because of the time difference. A man with a strong accent answered the telephone. I discussed the quality issue in detail with him. The valves were not sealing, and I questioned whether his test machine was functioning properly. I kept asking him when the last time was that they calibrated their test machine that applies a vacuum to test the seal on the part. He apparently could not understand me so he put someone else on the telephone. When I asked him about the vacuum test, the man responded, "Ah yes, suck test by mouth." I replied, "Say what?" "Yes, suck test by mouth. We testa every valva." Turns out they had cut a hole in a board, fastened a garden hose to the hole, and someone placed each valve over the hole and sucked real hard on the hose. If there was resistance, they passed the valve! Now this is manufacturing! True story.

My passion for manufacturing has been consistent throughout my career. I truly have been fortunate to have the enjoyment I have experienced during in my career. My energy and enthusiasm have directed my way more than once. During my time at OMC, they were executing quite a series of strategic projects. I volunteered for every project I could participate in. These were exciting projects, and I learned so much from them.

Lessons Learned

- It does not have to be new or modern to be Lean: an old galvanized steel building contained well-maintained equipment arranged in work cells with standard work and quality specs in place, and 5S was deployed. It was very much Lean.

- Passion, manufacturing, and the quest for excellence go hand in hand.
- If someone is yelling at you in a foreign language, they are probably mad at you. :)
- Problem solving is a skill, and all employees should be trained in it.

MULTI-PLANT INVENTORY REDUCTION

OMC had several plants in the Marine power products group during my time there.

> OMC – Waukegan, Illinois: Die casting
> OMC – Milwaukee, Wisconsin: Forgings, crank shafts, propellers
> OMC – Burnsville, North Carolina: Power head assembly
> OMC – Andrews, North Carolina: Lower unit assembly
> OMC – Spruce Pine, North Carolina: Lost foam casting
> OMC – Mexico: Harness, carburetion
> OMC – Hong Kong: Harness, components
> OMC – Calhoun, Georgia: Final assembly
> OMC – Brazil: Engine assembly

By the way, take another look at the above list. Is that a *Gemba*, or what? Die casting, machining, transfer lines, heat treating, forging, vibratory debur, milling, assembly, lost foam casting, annealing, lost wax casting, painting, you name it. I get sad when I think that it is all gone. That was an amazing company with amazing people and product for a long time. [I just noticed that my spell-checker does not have "debur" in it. I had to add it to the dictionary. What is happening to manufacturing?]

In 1995, the company was struggling to control its raw materials and work-in-process (WIP) inventory levels. It was straining the company's working capital and the CEO wanted something done. The company engaged the services of Computer Sciences Corporation (CSC) and charged them with leading an initiative to reduce inventory across all plants. I was asked to be the purchasing representative on the team.

This situation was ugly and political. Blame was being assigned everywhere. The plants were pointing the finger at purchasing, complaining that suppliers were shipping too much inventory, and shipping it too soon.

Purchasing was blaming the plants for not managing their schedules. A lot of ugly and negative energy was in the air over this situation.

CSC sent their team in and we had an initial meeting. They were all British—with the accents and all. I really hit it off with these guys. They were sharp, shared my passion for manufacturing, and liked to drink beer after work. I have always enjoyed working with the British. We would be home-basing the project out of the Calhoun final assembly plant. I would be flying out on Sunday afternoons and returning on Friday evenings. I was totally excited about being on this project.

We all met in the conference room at the Calhoun plant. The materials manager for the plant was on the project team. He would be participating on the team with us, along with a couple of other people from the other plants. He was a smart and an unbelievably quick-thinking manager. He did not like purchasing, however, and he never hesitated to bang on them when he could. To him, all roads for all problems ended in purchasing, and he would just crow and crow about it. However, he was wrong, and that would show at the end of the project. A couple of guys from product planning and the cross-dock function were also on the team. OMC had done a core competency analysis, and logistics was at the top. You literally could go to any computer, in any facility, and find out where a part was—on a truck, at a dock, on a shelf, at a machine being made, you name it. The accuracy of the inventory was uncanny. These guys were good. No, they were great!

The Calhoun plant accomplished the final assembly of Johnson® and Evinrude® outboards. These engines were part of the American culture and were fantastic products that were part of an exciting story. This factory assembled and shipped an average of 1,000 engines per day, and it was quite the place. It was state-of-the-art, and there was a whole lot going on in it. Power heads and lower units came in from other OMC plants, and components came in from the supplier. There were lower units, power heads, and cowlings passing around on an overhead conveyor system. In the final assembly work cells, there was a TV-type monitor that displayed standard work and suggested tools for the assembler to follow and use. In the center of the plant there were test cells with water tanks. Every engine was tested before shipment. The receiving docks were a flurry of activity…1,000 power heads a day; 1,000 lower gear cases per day: components for 1,000 engines every day…you do the math! At each dock was an expeditor. They controlled the flow of parts into the plant. They constantly looked at inventory reports and called suppliers to remind them of a shipment being due or asked them to move out parts because of a product mix change or other situation. These guys were all over it.

Lesson Learned

- The lesson I learned in this situation was how much wasted energy goes into bickering, blame, politics, and root cause speculation regarding a problem or crisis in companies. It also erodes the culture and the working environment. It is imperative to immediately put that energy into developing root cause analysis, Pareto analysis, affinity, and the tools of Lean Six Sigma. Speculation and politics are terrible tools for determining root cause and corrective action. Could you imagine creating a team and telling them, "Team, today's tool is politics. We are going to break up into groups of two and go to the corners of the room and take shots at the other groups while wildly speculating what the root cause is!" I think not.

BEARING SHORTAGE

Producing 1,000 engines per day was quite a marvel. Lots of things had to be where they needed to be, and be right, or it all went downhill fast. There was always a heightened sense of urgency in the air. Every problem had the potential to be a big problem. Call me weird, but that made it fun for me. It made my job exciting. I purchased all bearings for the company. Crank pin bearings, needle roller bearings, tapered roller bearings, ball bearings, plain bearings—all bearings. We purchased some plain bearings that were plated, etc., from a manufacturer in (I think) Greensburg, Indiana. These bearings were used on one of the smaller engines and were installed into the power head in the OMC – Burnsville, North Carolina, plant.

I remember coming into the office bright and early one morning. My telephone rang early and it was Linda Thomas, the expeditor for the Burnsville plant. She said they had a line down because the supplier did not deliver a shipment of bearings and were noncommittal on when they would ship. Evidently, one or more of their plating cells was down and they were behind schedule. It was now about 7:00 a.m. I advised my boss of the situation. He told me to go immediately to the supplier now and stay there until we were getting bearings. They contacted the company pilots to take me. I drove home, packed an overnight bag, and was at the local airport by 8:30 a.m. Moments later, two corporate pilots were flying me on the company jet to the supplier. At 9:30 a.m., I was in Greensburg, Indiana! Is that exciting, or what? Putting a buyer on a supplier site with issues was a common strategy for OMC. Some suppliers did not have the maturity to solve problems at the most efficient level, so we often assisted them. Another benefit was ensuring that the company was getting its fair share of the supplier's capacity. Having someone on site was very effective.

Sure enough, one of the plating tanks the supplier used to plate the bearings was down, and they were backed up. They were root causing and working to get it operational. They showed me our product awaiting plating. They said it would be a while. To pass the time, and because they were my supplier, I decided to do an abbreviated quality system audit and they accommodated me. I reviewed document control and their gage calibration program. Actually, they were doing quite well. They said it would be a while until we would get product. I advised them that I was told to "live there" until we got product. Well, we negotiated and they were able to commit to shipping a small lot of product, which was enough to ship the current engines in production. Production control was able to shift some other orders around in the schedule to buy us some time for the supplier to get back on track. They made the commitment, and I was back on the jet on my way home.

Lesson Learned

- Just as there is no substitute for "go and see" in the Gemba to understand or resolve issues, it is the same with suppliers. Within reason, getting someone on site to "go and see" during a supply crisis is highly effective.

STICKLE BRICKS: ONE-PIECE FLOW

OK, back to the conference room. So the CSC team gave us an overview of the project, milestones, and our schedule. It was most likely going to last several months, and we would travel to several of the facilities. During the overview, they communicated that we would be using certain tools such as process mapping and brainstorming, and additionally we would be doing a "one-piece flow" exercise with something they called "Stickle bricks™," which are the British version of Legos®. This was my first exposure to one-piece flow skits and exercises; and to this day, it always amazes me when they show the reductions in WIP, cycle time, and increases in on-time delivery. I think we should all participate in these exercises twice a year, literally.

All this Just in Time stuff was really catching my interest. I had also recently attended a three-day Just in Time training course with APICS in Chicago. To this day, it was some of the best instruction I have had in my career. The instructors were passionate, high energy, funny, and very well versed in the Just in Time philosophy. They kept our attention, and there was never a dull moment. This training had really awakened my interest in all things Lean.

The consultants set up the conference room with informational sheets; piles of Stickle bricks, and moved tables around. Everyone on the project team and some of the plant staff were given an overview of the Acme Corporation, which manufactured widgets in three or four different configurations. Some very rough bills of material (BOMs) were passed out, order demand was given, and inventories for various parts and components were provided for each process station. People were assigned roles as salesmen, materials handlers, expeditors, assemblers, etc. This was the proverbial, yet very revealing, one-piece flow simulation. After a few cycles of production, there were high levels of WIP everywhere, defective units, and a sizeable backlog of late sales orders.

Over the next phases, Kanban, accurate BOMs, mixed-model scheduling, and quality at the source were introduced and over the phases, on-time delivery improved, quality improved, inventory levels dropped, and product cycle times were cut in half. It was so simple an exercise, yet a powerful presentation of Lean principles.

AFFINITY DIAGRAM

John, the materials manager, may have been a ruthless debater but he was an interesting sort and graciously offered the use of his home as a place to conduct the brainstorming sessions we would be doing to start off the project. So we took him up on it. The next day, we all met at John's house and went inside. His home was a medium-sized, older, Southern-style home.

One of John's hobbies happened to be collecting old, generally upholstered and padded, chairs, but they had to be really comfortable old chairs. He had about six of these chairs plus a couple of couches. We all grabbed a chair and immediately began to comment on how comfortable they were. They were almost too comfortable.

The guys from CSC put up two or three easels and put out some markers, pencils, and sticky notes. To this day, one of the things that impresses me most about applying Lean principles or using Lean tools is how simple they are and that you do not need a lot of expensive technology. Eyes, ears, "work boots," pencils, sticky notes, easels, and some markers are really all you need. And some energy.

The first approach we took to brainstorming was to construct an "Affinity Diagram." The leader explained that he was going to ask one or more questions, and we would go around the room one at a time and take turns giving every notion or thought we had that was an answer to the question. He said that there was no such thing as a bad or silly response. He asked that no one criticize or make comments about anyone else's response. He explained that this was not a discussion, and that we had to refrain from starting one. He said that it might be hard but that we would have plenty of time for discussion later. He then told us that we would go around the room until we ran out of responses. If it was our turn and we did not have a response, we should say "Pass," and we would move on to the next person. Once everyone said "Pass," we would move on to the next step in the process.

The leader then posed the question: "Why do we have excess inventory throughout the company?" We began going around the room.

- Because purchasing ships too much in
- Because the plants are making schedule and using it up
- Because suppliers are over-shipping us
- Because suppliers are shipping early to get an invoice out
- Because our inventory is inaccurate
- Because our reorder points are wrong
- And on and on…

Each time someone would say a cause, the leader wrote it down on a sticky note and stuck it up on the easel. This process went on for about two hours. I found this an interesting approach. In a way, it was also a little stressful. You could only give one idea at a time, and there was the constant temptation to respond to someone else's suggestion. It was a room full of critical thinkers, and they all wanted to be critical talkers. All in all, at this point we had well over a hundred sticky notes. We broke for lunch.

After lunch, the leader said that we were going to separate the suggested causes by category. We were going to "affinitize" them. He asked us to stand as a group in front of the easels and silently group all the notes into categories at will. There would be no talking by anyone in this stage. It was an interesting approach. There was a lot of activity and arms going back and forth, and mannerisms of "that goes here; wait a minute; no, it goes there." Team members could move things at will, and if there was a disagreement, a duplicate note would be created and it would be assigned to more than one category.

Next, he asked the team to develop "category" or "header" cards for each group. If one of the notes served that purpose, they would use that note as the category (header). Otherwise, a category was created with a phrase that described it clearly. Discussion was allowed during this step. Once the categories were created, the completed affinity diagram was drawn. It was topped off with a problem statement that was created by the group. In our case, it was excess inventory across the company. We spent the remainder of the afternoon discussing what was on the diagram. You could see the project team beginning to gain a better understanding of the excess inventory, and there were some differing viewpoints. We finished for the day. John had advised us that he had arranged for dinner to be catered at his home. We took a break, some people made some phone calls, and the food arrived. We had an enjoyable meal. What a pleasure it was to be in the company of professionals sharing a passion for manufacturing and improvement. I learned so much from these folks.

Just when we thought it could not get better, John brings out a box of pretty high-quality cigars and a bottle of scotch. OK. So we had some drinks and a good cigar. He then brought out what appeared to be gifts and set them on the buffet. Out came a deck of cards and poker chips, and he told everyone to have a seat. We played poker for over an hour, just for fun; and when were finished, the guys with the most chips got the gifts. That was a great night.

The next day we met in the conference room of the Calhoun plant. We proceeded to do cause-and-effect diagrams for each of the categories of excess inventory drivers. Man, materials, machines, methods, and any other category that might be needed. My recollection is that we ended up with four or five key drivers, and the leader assigned each one to groups of two to manage. The team would report to itself, keep the team updated, and manage the collaborative effort in developing corrective action for its category. We spent the entire morning on this effort and finished up with our associated assignments.

In the afternoon, we headed out to the factory floor to tour the Gemba. We started at the loading docks and walked every step of the factory. We observed and noted all inventory in all locations in the plant. There was a wide variety of inventory: palletized, containers, boxes, bins, and crates. We obtained a scaled factory layout diagram and annotated locations and types of inventory storage. We talked to members of the materials team, production supervisors, and operators to get any input we could from their perspective. Operators always have such a great view of the factory. They see the world so differently; and if you are smart enough to talk to them, you can gain valuable insight from them. We would be traveling to OMC – Andrews and OMC – Burnsville the next day.

Everyone agreed to meet at a local restaurant for dinner. I looked forward to it. I really enjoyed socializing with the guys from CSC. They had exciting jobs, but it was impossible for a married person. One of the managers explained to me that he had just come from a week in India, and after spending this week with us, he would be off to Brazil the following week. These were all places with projects they were heading up. What an incredible opportunity for a young professional to travel from place to place throughout the world and lead Lean activities.

Lesson Learned

- Lean activities or tools such as affinity, cause and effect, brainstorming, and go-and-see generate positive feelings and teamwork. They focus the energy on solving the problem.

THE WAR ROOM

While at dinner, I noticed a tall gentleman talking with some young girls. He reminded me of John Wayne. This man was in his sixties but carried himself well, and he was interacting with those young ladies like he was in his twenties. His zest and passion for life really impressed me. They were just laughing, carrying on, and having a good time. The next week when I returned to Waukegan, I learned that that man was John (I cannot remember his last name). He had been hired as a key consultant for the Ficht project. He was a quality professional. By the way, that was not a story about an old man chasing young women, because he was not. It was about a man living life to the fullest, and I admire that.

The Ficht project was OMC's effort at producing direct-injected two-stroke engines with increased fuel efficiency and in compliance with the new EPA (Environmental Protection Agency) regulations. The company had bet its future on this technology, and it was a tough product launch, to say the least. These engines were highly technical, computerized, and deployed an injector modeled from German ink jet technology. The prototypes ran almost silently, had significantly greater fuel efficiency, and had very low emissions. I had taken a test drive on a boat with a pair of these engines. You literally could not tell they were running at idle. However, this technology was neither durable nor reliable, and it was the final blow to OMC. The Ficht technology replaced carburetors with fuel injectors on each cylinder; oil was injected into the crankcase, and the engine was run by a programmed electronic control module. These were major changes to the configuration of a two-stroke engine. Engineers worked right down to shaping the cloud of fuel injected into the engine. However, several design challenges manifested in the design that presented significant challenges that the company ultimately never overcame. The concept of direct-injected two-stroke engines was legitimate but it required more development, and that was ultimately accomplished by the Bombardier Corporation.

John, the quality consultant, had an interesting background. He had worked for NASA. It was said that he was in a key quality role for the production of Apollo 13. Once a spacecraft launched, the supporting production staff took time off, as the preparation up to a launch was gruesome. John was evidently one of those people and had started a vacation right after a launch. As the story went, he was standing in the surf fishing somewhere in Florida and someone called out to him that he needed to come back. There was a problem with Apollo 13. Great story; I have no idea if it is true, but I believed it and still believe it today.

One of the actions John took was setting up a "war room." I had been in "tactical operations centers" (TOCs) in the Army, but not in business. This was an intense project with a lot of pressure. Remember: 1,000 engines per day pressure. I had the opportunity to attend some of these daily stand-up meetings in the war room. There were approximately eight to ten speakerphones on a table, one for each plant. Upper management was there, purchasing, and engineering representatives. Each plant sounded off with any issues they had, and the appropriate areas would respond or make note. Each engineering area would give a status report—pistons, injectors, electrical, software, etc. To this day, this was one of the most productive and best-run meetings I have ever attended. There was also some drama. A multi-billion dollar company's future was on the line, and all hands were on deck. It was tense. It was manufacturing's version of war, and this was combat.

Lesson Learned

- A "war room" and daily or periodic all-hands-on-deck meetings can be highly effective tools for coordinating the elements of a key or critical project

THE PROJECT CONTINUES...

The next day, we took a van to OMC – Andrews. I rode with a gentleman named Bob and another, Dennis. Along with a gentleman named Paul, they managed the Master manufacturing schedule and I believe the Cross-Dock function for OMC. These guys were experts and managed these areas with serious discipline. They presided over the flow of components required to produce 1,000 engines per day throughout a multi-plant environment. There was not much room for error. I learned a lot from these gentlemen.

There was lively discussion the whole way. Getting accepted at OMC, a very old-style manufacturing company, was not easy. The culture was thick. It was on this day, for some reason, that I was given one of the secret handshakes. I told some story, and these guys thought it was funny and they ragged on me about it the rest of the trip. It felt good to start to be accepted.

We toured the Andrews facility just as we had toured the Calhoun plant. We spent the majority of the day there. Being the only guy from purchasing around two OMC materials managers for a day made me nervous. These guys had tough jobs and they "ate" buyers for breakfast. As all OMC facilities, the Andrews facility was impressive, well organized, very orderly, and well run. The next day, we toured the Burnsville plant and then took the jet home. We ended a very challenging, but rewarding week.

We all did our homework assignments over the next couple of weeks and were scheduled to give upper management a progress update in Waukegan. One member from each two-man team presented. The project team was slowly moving toward a consensus that was assigning root causes as:

- Manual mailing of blanket purchasing orders and associated updates took several days to process and the suppliers could not respond quickly enough when inventory needed to be rescheduled to later dates.
- Some supplier lead-times were excessive, driving up larger order quantities and unnecessary commitments into the future.
- Buyers need to be more proactive in holding accountable those suppliers who over-shipped or shipped early.
- The final assembly plant in Calhoun needed to perform better to schedule.
- The plants were using too narrow an inventory management strategy and were not considering certain dynamics across different types of inventory.

Corrective actions:

- Switch as many suppliers as possible over to Electronic Data Interface (EDI) to provide significantly faster response times.
- Work with suppliers to reduce their lead-times, or switch to suppliers with shorter lead-times.
- Hold buyers accountable for holding suppliers accountable for not over-shipping or early shipping.
- Have buyers be more aggressive in rescheduling incoming shipments out when needed.
- Charge OMC – Calhoun with implementing a project to improve schedule attainment.
- Apply the runner, repeater, and stranger inventory approach to managing inventory items. (This entailed using an analysis to assign an inventory management strategy of MRP, Kanban, or two-bin system for all items, depending on how often they were ordered, what they cost, how many were used, etc.)

The project concluded with good results. OMC was able to reduce the average on-hand inventory across the company. I learned a lot and moved on to other things. This was my first key experience using some of the Lean tools to solve problems.

DESIGN FOR MANUFACTURE AND ASSEMBLY

Engineering announced that the company had asked CSC to lead a design for manufacture and assembly (DFMA) exercise for the Ficht engine. DFMA is a component of Lean manufacturing that is not talked about as much as some of the others, such as one-piece flow, 5S, poka-yoke, etc. The purpose of applying DFMA is to identify, quantify, and eliminate waste or inefficiency in a product design. The analysis is helpful during both the design and production phases of a product's life cycle, but more valuable during the design phase. DFMA will assist in creating products that are less costly and easier to produce. Additionally as important, the fewer components there are, the fewer there are that need to be designed. Developing fewer components shortens the product development time, and there should be less design changes once the product goes into production because there are fewer parts. This can be a highly beneficial tool.

They wanted a representative from purchasing to participate, so I volunteered. This was a great and interesting project. There were several engineers, engineering technicians, myself, accounting, and some materials folks. It was a solid team.

In preparation for the kick-off of the exercise, I and three other people were given the interesting task of "wrenching down" a new prototype engine. That's right; we went into a good-sized room with some tables and a V-6 Ficht engine on a stand, and a toolbox on the floor. We just started wrenching parts off of the engine and spreading the parts around the room. We started sorting them by component type—casting, stamping, forging, electrical, etc. We did not go to the nut-and-bolt level, but we broke down the engine into what I would call "chunks."

The team met in a conference room. There were a couple of computer terminals in the room with access to the BOM for the engine. The project leader gave us an overview of how the DFMA process worked and provided us with a practitioner's manual. I still have my copy today. I am not going to go into detail of how to conduct a DFMA exercise, but I will review the basics of the concept.

Basically, the process asks three things:

1. Does this part move relative to the part to which it is attached?
2. If the part is attached to a different part, does it need to be made of a different material than the part it is attached to?
3. Does this part need to be removable?

If none of these conditions occur, then the part is a candidate for elimination or combination with other part(s).

This project was laborious and lasted almost three weeks. We literally took every part and asked if it could be eliminated or combined with another part. We removed something like sixty-four parts from the engine. I was really fascinated by this result. I have mentally referred to this process many times since this experience.

PROVEN TECHNOLOGY VERSUS HIGH-TECH

One of the components I purchased were needle roller bearings for crankpins. The crankpin bearing goes in the flange at the bottom of a piston connecting rod that circles the crankshaft. All the energy from combustion is transferred through these bearings, and they live in an extremely violent environment. It is akin to living in the middle of a continuous explosion. This is a critical part and has some important product tolerance ranges for certain attributes of the individual roller needles. There is a limit to the size differences between each bearing and its companions, a limit to the tolerance stack between all of them, and to a condition called "bar skew." If a crankpin bearing fails, it is not good as it is almost certain to be catastrophic to the engine. All crankpin suppliers had large quantities of rollers stored by diameter ranges in groups. To meet the tolerance stacking specification, they would select and create sets of rollers that fell within the spec.

Crankpin bearings were purchased from two or three different major suppliers. There were a couple of different designs for the cage that holds the needle rollers. The traditional design used a cage that was punched from a thick piece of steel, and the other was stamped from a piece of steel. I conducted quality system audits at the factories that manufactured these bearings. We had one supplier in particular that had been around for a long time. They had modern, up-to-date equipment, and they also had some older equipment. One of the other suppliers was state-of-the-art "cutting edge." I visited their factories and was quite impressed with their modern technology. They had some fascinating transfer lines and automation, and their product quality was good. They had provided some competitive quotes and had submitted samples that passed some rigorous testing requirements. It was a different type cage and they really talked it up, and in the end it was a fine product.

However, the current supplier's bearings were functioning quite well with a favorable field failure rate. When I visited this supplier to do a quality system audit, I was immediately taken aback by the age of the equipment making our bearings. At first, I asked myself if that was acceptable. I mean, we were a solid customer. Did we not deserve to have our product produced on newer equipment? The quality survey revealed that although the equipment was old, it was well maintained, and it demonstrated a process capability well within what was required for the product. I made the decision to stick with this product, and they continued to supply these bearings for several more years. My point here is that I almost fell into the "sexiness of fancy," or "new," or "high-tech." In Lean, "simple works and they prefer proven technology." That is one of the more powerful beauties of the Lean philosophy. Lean principles recommend going with a proven technology whenever possible, not necessarily ruling out new technology, but not just for the sake of going with new. It is also much easier to get support in taking action that does not involve fancy software or expensive equipment.

Lesson Learned

- Be careful not to fall in love with "sexy" or new equipment. Lean prefers proven technology. If equipment is still within its useful life, capable, and delivering consistent quality product, it is most likely best to stick with it unless there is some significant benefit delivered by the new technology.

COMPETITIVE PRODUCT COMPARISON PROJECT: BOSTON CONSULTING GROUP

OMC continued on in its effort to reduce costs, improve product, and maintain a strong presence in the outboard market. They were interested in comparing the cost and components against their competitor's product. This was another fascinating project. The company prepared a large room, about the size of a wedding hall, and set up tables much the same. The room was cordoned off in two halves. On one side they were disassembling a Mercury V-6 outboard, and on the other side a Yamaha.

As the team disassembled the engines, they would bring it up to a person at a computer terminal who would name the part, describe the part, and enter in a vendor or manufacturer's information that may be indicated on the part. The teams constructed BOMs for each engine.

Once the components were cataloged, purchasing obtained quotes from our suppliers, many of them the actual suppliers of the components. My job was to take the bearings, cables, and some other components I purchased and go out into the market for quotes. Once quotes were received, they entered the pricing into the BOM and determined a pretty reliable estimate of what their chief competitor's cost structure was. The results were kept confidential, but I do remember that some of the components I purchased were of higher quality and cost than the competitor's, yet they were over-specified.

Lesson Learned

Specifying components far over the required capability or requirements can be very costly, make product unnecessarily over priced, and take dollars away from the bottom line.

9

Lean Transformation # 1

In 1997 in an inconspicuous factory on Northwood Drive in a small industrial park in Maryland, I began my first factory Lean transformation. By that time I had earned the APICS certification, including the Just In Time module certification test, and traveled to Japan and Taiwan conducting quality system audits and studying Japanese manufacturing techniques. I had read *Just in Time for America* by Henry Wantuck, *20 Keys to Workplace Improvement* by Iwao Kobayashi, and several other Lean or Just in Time books. I had attended demand flow technologies training and had gained some practical experience in applying some of the Lean tools, such as affinity, pareto, mapping, one-piece flow, and Kanban, line balancing and leveling, Takt times, Visual management, Andon, and Jidoka. And now, I had my very own Gemba!

One evening a few months later, I received a call from the director of purchasing at a large boat manufacturing company that manufactured several popular brands. He introduced himself and said that someone had mentioned my name to him and indicated that I was interested in being a plant manager. I told him yes. He was quite the friendly sort and we had an enjoyable conversation. He didn't mention any specifics, but he suggested that I continue to pursue joining their team.

A couple of weeks later, I received a call from the vice president of operations who was looking to hire a plant manager for one of their factories in Maryland. I ended up taking the job. What was exciting was that I more than doubled my base salary, and about twelve months later I got a bonus that was equal to my annual salary as a technical buyer. That was fun! As excited as I was to become a plant manager, I was sad about leaving OMC. It was such a great company with such great people. The knowledge I gained about manufacturing and the opportunity to attend training while at OMC had a significant impact on my career for which I will always be grateful. OMC was to go out of business and cease to exist in the future, but it certainly would not be a reflection on the people who worked there.

So one day I boarded a plane at O'Hare International Airport and headed for Maryland. My family stayed behind until I could get settled in and we could find a place to live, sell our home, and all of that. I arrived at the hotel and settled into my room. I was pretty wound up and did not sleep well. I headed down to the lobby at about 5:30 a.m. and grabbed a cup of coffee and a bagel. My palms were sweating. At about 6:00 a.m., Jim, the regional plant manager, met me in the lobby of the hotel. "Hi, I'm Jim," he said. "Welcome! Are you ready?" I told him I was. He said, "OK. I am going to take you to the plant and introduce you to your team and then I'm outta here." On the way, he told me that the plant had about 400 employees and was behind production for the month.

THE NEW PLANT MANAGER

We walked into the front office and into a large office. He called a meeting with the managers and supervisors and introduced me, "This is Bill; he is your new plant manager. Please give him your full support." Then he left. I was talking to my new staff, and for the first time in my life, I was a full-time plant manager of a boat plant! It was awesome. I would be less than honest if I said I was not a little intimidated. But this was my life's dream, and I was relishing it. I wanted to call everyone I knew and tell them I was a plant manager. I probably did.

This plant should not have been producing boats. In fact, they were not. They were just producing warranty claims. The models ranged from twenty-nine feet to forty-six feet. The flagship line, the forty-six-foot product, had about seven boats in assembly, two in lamination, and one in the test and ship area. They had an order bank with no end, and the plant could only make about one per month, and that one was not of very high quality. Corporate wanted boats shipped so they could make their numbers, so there was a lot of pressure, and I suspect some bonuses were on the line. I got a call a day from an angry dealer called to say, "Do you have a quality manager?" "Yes, we do" "What's his name, Stevie Wonder?" Was that necessary? Then the sales guys would call: "Man, I can sell every one you can build. I need them bad. I have dealers screaming."

On about my third day at the plant, I got a call from my boss one morning early. He said, "There's an engineer in your plant and he has a stack of pictures of all kinds of things. You need to find him and get them from him before he brings them back to corporate." I located the gentleman and recovered the photos. This was just inappropriate.

This brand, which I will refer to as Brand X, was an ego-driven effort to compete with one of the company's sister companies. This attempt failed miserably and, furthermore, should have never happened. The company manufacturing Brand X was a small boat company and lacked the required design for manufacture and assembly skills required to efficiently build larger boats, at least the express cruisers. What was crazy is that when it came to short product cycle times and Lean operations, the company was ahead of its time. It assembled small boats by the thousands and managed costs quite well. The company had developed a great deal of discipline in their manufacturing operations, but ego got in the way.

One thing unique about this company was that in some plants they assembled small boats in stations as opposed to the assembly lines that most, if not all, other builders used. Once the boats' fiberglass parts were moved into the assembly departments, they didn't move until they were completed. On one occasion, I had a lively debate with another plant manager who came up through the ranks on which was better assembly lines or station building boats. We had a lively debate and it was fun. As far as small boats go, I have formed the opinion that at best, there are pros and cons to each way. It was fascinating and I learned so much from the debate.

The Brand X, although stylish, were terribly designed and were almost impossible to build in volume or at low cost. The designs were filled with unnecessary obstacles to assemble. For instance, the forty to forty-six-foot Maxum hulls had a taper on the back of the boat. Engineering had designed them with a removable "bustle" that built in hours of waste and rework, plus generated finish issues later on in the life of the product. Sea Ray, on the other hand, had achieved the same style feature by applying a milder taper to the part. They simply lifted the bow out of the hull mold first, tipped it up, and pulled the part forward. It was a tapered part with no built-in rework. The company had designed most of the boats with bolt-on mold inserts to achieve a louver vent. These large bustle mold components had to be lifted up and bolted onto the hull mold. The seams between the bustle mold and the hull mold had to be filled with wax to minimize repair work for this built-in defect. Once the hull was completed, the bustle molds had to be removed in order to pull the hull from the mold. Even when filler wax is applied best, it still generates textures and prints on the finished parts that require fiberglass rework. This was absolute and totally unnecessary—not to mention the cost of creating and maintaining the bustle mold add-ons. This was absolute waste.

I also believe the company just viewed sport yachts as "big little boats." operators and plant workers can work around minor design issues on smaller boats, but the greater the size of the boat, the more important quality or fit and finish becomes. The design and fit issues on the bigger boats stifled the production process in larger boats. That combined with the constant unrealistic pressure to "make rate" ultimately generated high levels of warranty claims and poor-quality boats.

Further exacerbating this situation was that there was an absolute arrogance in design engineering and corporate at the time. It was bad enough that the product line was so ripe with quality issues, but what made it worse was instead of listening to the folks in the plant building the boats, they treated them with disdain and outright disrespect. To have people from corporate come into the plant and walk around snapping pictures of defects created by boat builders and openly criticize them was so unacceptable. Instead of asking the boat builders what they could do to help or make the product easier to build, they walked around pointing out discrepancies. The boat builders assembling these boats had so much invaluable knowledge about the design deficiencies, but they were not being listened to and so the design issues continued. Ultimately, there was progress made, but it was too little, too late.

The company continuously shifted production rates in the yacht plants just as they did in the small boat plants. They did not understand the impact of production rate changes within plants producing larger boats with longer cycle times and more complex work-in-process (WIP), especially in assembly lines. The range of skills is greater in large boat plants. The engine room takes one week for an operator to complete, including installing the engines and generator for a forty-foot express cruiser, yet an engine and associated components in a small boat can be installed in a few hours. The company did not understand the dynamics, importance, and in fact the requirement to consider and manage WIP footprints in plants producing larger boats. For example, let's say the assembly cycle time for a forty-six-foot sport yacht is fifteen days (120 hours) and the Takt time for the line is five days (forty hours), which requires three boats in the assembly line (3 × 5 days = 15 days). The company decides that sales support an increase in the Takt time to about six boats per month, or twenty-four hours. This decision requires the plant to increase the WIP footprint in the line from three stations to five. This means that the plant has to hire the people and build two additional units to establish the required footprint. Required WIP is only created in two ways: building more than you are completing, or completing less than you are building.

Given that the product cycle time (critical path) in assembly was fifteen days, it would take a minimum of thirty days to increase the assembly WIP to five boats, assuming labor could be shifted from other areas or hired, and additional raw materials and components could be obtained. Additional labor and materials would also be required in the fiberglass lamination areas and sub-assembly departments. Upper management did not understand this dynamic and imposed or pressured unrealistic rate changes on the plant. These decisions, combined with serious design deficiencies and a "focus on the numbers," drove high levels of stress into the plant. Employees were required to work unacceptable levels of overtime, they got tired, quality issues increased, and the plant did not deliver on time. This was a very caustic and less-than-fun environment.

The Incentive System

The company had an incentive pay system they developed in the small boat plants. Basically, employees were paid a base hourly rate, say $7.00 per hour. The company would pay the plant something like $16 per standard hour to build the boat. So if the boat was awarded 112 standard hours by the industrial engineering department, the plant received $1,792 for each of that model completed for the month. The employees were paid their base rate plus the difference between the total base rate payout for the month and the total standard payouts for all boats completed for the month. See Table 9.1.

Table 9.1 shows in Scenario 1 that with no overtime worked, an employee's effective hourly rate for the month would be $20.07 per hour. If all was the same, but the plant had only twenty-six employees and got the same boats done with zero overtime, the employee's hourly rate for the month would be $22.08 per hour (Scenario 2). Scenario 3 shows what happens if the employees worked $15,000 of overtime for the month; they would have made only $16.94 per hour. Scenario 4 shows the impact of completing fewer boats with the same employees with an hourly rate for the month of $16.71.

TABLE 9.1

Incentive Pay System

	Scenario 1	Scenario 2	Scenario 3	Scenario 4
Base hourly rate	$7.00	$7.00	$7.00	$7.00
Standard hourly rate for boat	$16.00	$16.00	$16.00	$16.00
Boat standard hours (assume 1 model)	112	112	112	112
Dollars plant receives for each completed boat	$1,792	$1,792	$1,792	$1,792
Boats completed for the month	35	35	35	26
Standard labor dollars paid to plant for month	$62,720	$62,720	$62,720	$46,592
Overtime dollars	$0	$0	($15,000)	$0
Standard labor hour payout minus overtime	$62,720	$62,720	$47,720	$46,592
Number of employees in the plant	30	26	30	30
Incentive payout per employee	$2,091	$2,412	$1,591	$1,553
Hours employee worked during month	160	160	160	160
Base hourly pay for the month (160 hrs × $7.00/hr)	$1,120	$1,120	$1,120	$1,120
Incentive payout for the month	$2,091	$2,412	$1,591	$1,553
Total employee compensation for the month	$3,211	$3,532	$2,711	$2,673
Effective employee hourly rate for the month	$20.07	$22.08	$16.94	$16.71

I am sure your mind is already turning. Lots of pros and cons with this system.

The pros:

- Employees hold their co-workers accountable for pulling their own weight.
- Employees hold their co-workers accountable for not working overtime.
- Employees often actually discourage hiring more people with turnover as the incentive money is divided among fewer employees effectively increasing their hourly rate.
- Product or process improvement is encouraged because, as a policy, the standard hours for the boat are never reduced.

The cons:

- Emphasis can easily shift toward getting product completed and away from quality.
- Rate changes both ways impact employees' pay and can discourage them.
- The company actually enjoyed a considerable amount of success with this pay system but, in my opinion, it impacted quality in a negative manner.

Significant problems arose in applying this to larger boats over thirty feet in length; and the larger the boat, the more the problems. The standard hours allowed for the models in the plant assumed a properly designed boat, and certainly could not have come from an industrial engineering (IE) time study, or they would have reflected the work-around and rework hours driven by the design issues. So, the employees in the Brand X plant basically had no chance of ever receiving anything over their base wage, and what made matters worse was they continually heard of the high incentive payouts in the other plants. I must have answered a million employee questions about an incentive program in a plant that never to this day received an incentive payout. In fact, I used some leadership discretion and announced that the program was discontinued. I was chastised from the top a little, but in the end the decision was accepted.

In addition to the inaccurate standard hours, the dynamics of rate changes and new product introductions in the bigger boat plants were too great to make the system feasible. It just resulted in tired workers, missed schedules, and terrible quality. Corporate would make terrible scheduling decisions and rate changes, and then would put serious pressure on the sport yacht plant managers to make rate. This was less than fun. They further had weak product development and did not hold them accountable for quality. The designs were lacking, the documentation and prints were inaccurate or inconsistent, and they did not understand the learning curve associated with large boats. It was all a recipe for failure. As I write this book, Brand X boats are no longer produced.

QUALITY AT THE END OF THE LINE

Each plant had a quality manager and a team of quality auditors (QA). The responsibility for product quality was assigned to the vice president of Human Resources, who was hundreds of miles, or in some cases thousands of miles away from the facilities. The company did not have a clearly defined quality process. There were very few product standards for large boats, and what standards did exist were not maintained. They did have a usable set of manufacturing standards for smaller boats. The quality manager and auditors were found on the end of the process. They would go through a boat and write down a list of everything they could find wrong. Hundreds of gigs or defects were identified. There were pages and pages, and list after list. Sometimes there would be bravado, such as QA walking off the boat because they were finding too many defects. What was even crazier was that there was little if any corrective action being taken.

Periodically, the vice president of Human Resources would get fired up and "jack up" the QA team, and for a few weeks the number of defects written up would triple. After plant visits by the VP of HR, the lists would quad-triple. What was crazy was that the VP of HR's office was about fifty feet from the design engineering department that created all the design defects. What a mess! It took me a year, but I finally arranged for the quality manager and team to report to me. However, it was not so much about reporting to me, as it was to have the quality team be part of the plant team and changing their role to helping get the product right as opposed to pointing out what was wrong.

The biggest dealer in Michigan was managed by a husband-and-wife team. They were tough, and they were not happy about the quality. These boats sold for $200,000 to $400,000, and the owners were not happy with bad quality. It was legitimate. They had received boats with unbelievably bad quality. The wife, in particular, was tough as nails. She was brutal on bad quality, and it was nearly impossible to make her smile—and I didn't blame her. To make matters worse, during a recent plant visit by her, a boat builder had drawn a large cartoon of the male anatomy on the back of a bulkhead that she was able to see while walking through a boat. Great, just great.

The tension at this dealer continued. It was stressful. Initially, the harder we tried, the worse we did. I would personally go through every boat with a flashlight. I caught a lot of things, but there were so many that the statistics were against me. Every time we would ship a boat, a couple of days later the phone calls would start, either from someone at corporate or from the dealers themselves. They would write up lists of defects and issues, and fax them to us. I could hear the fax outside my office printing page after page, after page. Man, did I hate the sound of that fax machine!

As it continued, I was summoned to go visit this dealer. The wife of the team met us at the door. She was "all business." So she took us through the boats and pointed out defect after defect. She would say, "Look at this. How could they let this go?" It might have been a mis-drilled hole or a door missing screws, or a scratch, or a light that had no wires behind it. Some of the defects were horrendous and embarrassing. It was clearly the result of getting boats out, as opposed getting them right. While I was looking in the bilge in one boat, I noticed what appeared to be the absence of a mounting block of wood under a bilge pump fastened to the bottom of the boat. When I got out of the boat, I looked under the boat and saw screws protruding through the bottom of the boat. Mind you, this boat sold for over $300,000.

The company was applying wood and synthetic wood veneers to plywood, then staining them, and covering cabinets with them in the boats. For the price, they had a good look. However, there was a quality issue with the stain or the materials being used. Customer complaints and warranty claims were starting to come in, and I traveled to angry dealers to survey the issues. A few months after arriving, in about February I traveled to a boatyard in Martha's Vineyard to review a boat with interior wood issues. It must have only been about 15 degrees outside. We crawled up on an icy boat in a boatyard, cut a hole in the shrink-wrap, and crawled on our hands and knees to the entry door. The boat was a mess, and we made the decision to bring it back to the plant for repairs. The repairs required pulling all the cabinetry out of the boat, replacing the veneer, and refinishing it. "Un-building" a boat is no easy task, and something no plant manager wants to do. One thing I have noticed over the years is that a unique dynamic sometimes occurs with customers. Customers who have a major problem with their boat with a very expensive warranty claim sometimes are more excited about their boat and the company than some who have very little go wrong with their boat. It is weird, but taking responsibility for a big defect that should not have happened can build a greater level of loyalty from the customer than one who owned a boat with few or no issues.

To construct the curvy Euro-type styling so popular in express cruisers, the industry used a lot of Luan, or "bender board." This is a wood that is very pliable and can be wrapped or bent around curvy structures to achieve smart styling. Well, the story goes that Luan is cut in the forests of the Amazon and other jungles. The logs are floated down the rivers to the mills. We started to get calls from dealers that customers were complaining that their boats smelled like "vomit." You cannot make up this stuff. I flew up to our biggest dealer again, and reviewed one of these situations. Sure enough, a nearly half-million-dollar boat smelled like vomit. You have to be kidding, I thought.

Although, I really do not know for certain to this day, supposedly some Luan logs spent too much time in the river before arriving at the mill. When logs were in the river too long, they filled up with bacteria, which ultimately over time began to smell…like vomit. So, away we went. We had to send dealers replacement pieces of upholstery panel with new wood, and in some cases had to travel to boats to replace wood and do other things to remove the smell from the boats.

THE AFT CABIN

The years 1997 and 1998 had very good markets. We had a lot of orders for product and frankly the sales were limited only by the plants' abilities to produce (at least that is what sales said). The company had two runabout plants a couple of hundred miles away. Well, our plant was producing a forty-foot aft cabin boat. This boat was a nightmare to build and was the opposite of manufacture for assembly. There were a lot of orders and dealers were screaming. Management set up a plan to make all the components in our plant, ship them on a truck to a runabout plant a couple of hundred miles away, and assemble these forty-foot boats there. Anyone with marine industry experience will tell you that the skill sets and work mentality of boat builders vary considerably between runabouts, cruisers, sport yachts, and yachts. The bigger the boat the slower the pace as there is more work content, and the higher expectation for workmanship and quality. Not to say quality is not important in runabout plants, but the work pace is fast and it is much simpler work. A runabout plant that builds 100+ boats per week is no place to build a sport yacht and vice versa. It's kind of like having a veterinarian do open heart surgery on a human. Both require talent, but they are just different.

As someone who worked at both this company and its sister company, I noticed some big differences. The primary value at the sister company was quality, quality, and quality. Everything had to be right. Period. At this company, the producer of Brand X, the primary value was "making rate." Once a week you would hear someone say, "We gotta make rate." An overemphasis on volume as opposed to quality combined with very poor management decisions and ego greatly impacted this company. In my eyes, the company producing Brand X focused on "getting things done," and its sister company focused on "getting things right." That to me is the primary reason that the sister company has survived and as of today the Brand X company has faded.

In a plant that was already struggling from years of bad management and assembling boats with significant design-for-build defects, workers were being asked to box up components highly susceptible to damage and ship them for assembly in a plant not experienced or equipped for assembling large boats. This was a huge distraction that just added to the already-burdened plant, not to mention the costs associated with this nonsense.

It was a never-ending flow of damaged parts from shipping, rework, and disruption. Forty-foot boats need to be water tested in a test tank and preferably run on open water before shipping them to a customer. The runabout plant had no test tanks. So they plugged the drain in the loading docks, filled them with water, and floated the boats for testing. It was a fiasco. I disagreed with this activity in every way. We may have been making boats, but there was certainly no money being made…other than maybe someone getting a bonus.

Although two employees carefully accounted for each part by touching them and marking them off on the shipping list, the runabout plant would call and complain that the boxes of components were not complete and they needed more parts. Our plant would have to immediately respond and produce the supposedly missing components. This disrupted the plant even more. Later, the "missing" parts would be found. What a mess. You could write a book about this. Oh, wait a minute. I am.

Let's take a breath here for minute … the VP of HR/quality sits fifty feet from design engineering, who is proliferating poorly designed boats one after the other. Lists of 200 to 300 defects are being generated at the end of the production process as fast as the QA team can write them, warranty is out of control, the plant is sending people to dealerships to fix or repair another fifty defects per boat, and the people at corporate are running up and down the halls criticizing the plant for poor performance? Welcome to manufacturing in America in the 1990s!

The sales team is paid big bonuses for shipping product to the dealer and the dollars of product to dealers. Each dealer has a credit line. So, the sales bonus program rewards sales guys to stuff the dealer with product they may or may not need, and may just well put them out of business. The rest of the company's bonuses are right in line behind it. The dealer programs also incentivize the dealers to take more product, giving them volume discounts. Another problem was that in this environment, demand would halt quickly. All of a sudden, dealers would say, "Enough!" This was very disruptive to the company. Although everyone in this environment for the most part was working hard to do the job they were asked to do, they really did not see the forest for the trees. In fairness, one of the new presidents of the company created a "pipeline report" and also created a focus on inventory throughout the pipeline. This was sorely needed.

So now the whole company is in the race. Are we making rate? Are we shipping boats? Where are we for the month? Where are we for the quarter? We have to make the "numbers." Hire more people if you need, but make the numbers...people are hired, resources are spent training them, and there are cost variances driven from hiring. Then the dealers choke and say, "No, no, no. We are not taking more product. Enough is enough, no matter what the discount is." OK plants, all those people you just hired away from other employers and trained, lay them off now.

RATE VERSUS QUALITY

In a discussion one day with someone from corporate, I brought up my curiosity about why the quality team would not report to the plant manager as next to safety; quality was the manager's number-one responsibility. This person said, "I think the company knows that plant managers are primarily worried about making rate and getting boats out the door, so they put quality under HR." Somebody help me, please.

I asked this person, "What do you think the biggest cause of not making rate is?" "I'm not really sure," he said. "Do you think it's because people just don't care?" I asked. "No, I mean everyone seems to be working hard. It just seems that there are too many things wrong with the boats at the end of the process." "I asked him, "OK, so why do you think that is?" "People are in too big of a hurry to get the boats done." "Oh," I said. "Well don't you think having to go back and rework all of their quality defects slows them down?" "Well," he said, "they aren't the ones doing the rework. The people on the end of the line are. It takes too much time to find everyone and bring them to the end of the line to fix things. It's easier for the people at the end to just fix things themselves."

"Well, how are they going to learn how to do it right if someone doesn't tell them it's wrong, or how it's supposed to be?" "That's a good question," he said.

I suggested to him that in my experience, the number-one reason plants do not achieve production schedule is the amount of quality defects and associated rework that occur during the process. If every part leaving a department showed up defect-free at the next department, product would literally fly through the process. Therefore, if a manufacturing plant manager truly wants to be successful at "making rate," he or she must place a huge emphasis on total quality throughout all processes and at every stage. That just takes you right to the thought that if quality is so crucial to a factory's successful performance, then it only makes sense for the plant manager to manage those resources.

Let's Summarize

- End-of-line defects were numbering in the several hundreds.
- All quality resources focused on inspecting product at the end of the line.
- Very few product quality standards existed, were maintained, or were used. A defect was anything a quality auditor decided it should be.
- The quality auditors were often hired off the street with no formal quality experience or boat product or process knowledge. There were quality auditors who literally did not know the bow from the stern, or have the slightest understanding of boat building, or quality for that matter.
- The quality function was assigned to the VP of HR as an extra duty because a plant manager could not be trusted.
- Upstream operators were not being made aware of their defects, so they continued to produce them.
- The VP of HR/quality managed by fear, driving the QA team to audit product harder and harder. The level of product rework was affecting the plant production throughput.
- There was no formal quality or corrective action process.
- Design engineering was designing rework right into the product, such as add-on bustles requiring hours of built-in fiberglass finish repair.
- There was so much rework on the finished product that it looked like it was three years old when it left.

- The wage incentive focused solely on the quantity of parts and product.
- Although the quality resources reported to the VP of HR/quality, the factory managers were held accountable for the quality of the product coming from the plants. Because the quality resources were not under his control, he could not refocus them from the end of the line to resolve the issues at the point of creation and train operators in prevention. In the vast majority of cases, the operators upstream in the process were not aware of the defects they were producing, as there was no effective feedback.

Are we having fun yet?

This was a tough situation for me because this thinking was deeply engrained within the culture. Being new to the organization and challenging this situation just made me look like I was not adjusting to the culture. And I wasn't. I was going to change it. I had to or my plant would fail, and I was not going to let these people down, and failing for me personally was not an option.

THE LOOP

One day on a conference call, it was announced that we were having a "loop" in a couple of weeks. I asked what that was. The president, his staff, and some of their team would be flying in and visiting the plant. Representatives from engineering, quality, and HR would be walking through the plant and checking on things while the plant manager did question-and-answer with the president, the VP of operations, and other members of the corporate staff. After a few hours, we were scheduled to gather in the conference room for an out-brief. Each member would advise of everything the plant was found to be "doing wrong" and provide a list of all the things they found.

The executive team would travel via the corporate jet. I headed to the local airport to pick them up. I was a little nervous, as having the entire corporate staff in your plant in the middle of a turn-around can intimidate the best of us. On the other hand, as big as it was, I was excited about the challenge and the vision for the factory that I had in my head. I had now had all of about three months of experience as a plant manager.

It was quite a sight to see the Lear jet do a half-circle of the airport and land on the small runway. They taxied up and I met them as the door opened. They were an uneasy-looking bunch as they came down the steps of the jet. We exchanged a couple of handshakes and formalities and got into the van. As I drove back to the plant, not much was said. I made a couple of comments about how we were looking forward to their visit. They did not give much of a response. A couple of comments were about the scenery and a couple of basic questions to me about employee head-count, etc., but that was about it. OK, onward and upward.

Everyone headed toward the plant's conference room, and immediately they asked where telephones were that they could use. A few underlings headed right for the shop floor, and the rest talked on their cell or company phones. I was immediately thinking to myself, "OK, they flew 3,000 miles to talk on the telephone?" But, I was a factory guy. What did I know?

I greeted my boss. He was a good guy, and I really appreciated him. He could see in my eyes that this event was strange to me. He said something to the effect of, "Just roll with this, it will be over soon." One by one, the telephone conversations were ending for most, but not all. As the president ended his conversation, I welcomed him and said something to the effect that, "Sir, the team is working really hard to fix this factory." He said, "This factory is losing money, and frankly I am not sure I'm not going to close it down." He was really negative, and his conduct did not impress me.

I was thinking a few things. One, I hope he did not hire me away from a secure job with a great company to put me out on the street after he closes this plant. Two, I was thinking that he did not seem to understand that this factory deteriorated to a terrible state of performance under his leadership and that he might want to take some responsibility. Three, what it would take in effort to fix this factory was something no normal person would sign up for and that he better start appreciating my presence and efforts quickly. He really aggravated me.

This man had a strange personality and not what you would expect from a company president. Some people had told me that he was "really religious and not to cuss in front of him. I really did not cuss that much, so it was not a big deal to me. On the other hand, this man did not reflect any kind of religion I was familiar with. One of his rants and themes during the visit was "integrity," and he defined it as doing what you said you would do, and doing what you said you would do was to "make rate." He never once mentioned "quality." In reality, it worked like this: He and his team would pound the plant managers for production plans required to earn the executives bonuses. They would pressure and pound, and come at you from every direction until you finally committed to a plan that was impossible to achieve. The plants would be over-scheduled, and then add to that the quality situation. What a recipe for an exercise in futility. I certainly did not think that not achieving a plan that was absolutely shoved down your throat indicated any lack of integrity.

I asked the president if he was interested in touring the plant and seeing the shop floor. He was not interested. Unbelievable. We were a manufacturing company. The day continued to progress with a lot of the group talking on telephones, etc. Everyone started to progress into the conference room. Different members of the group got up and reviewed their findings, basically a laundry list of everything they saw that was wrong. I myself had a fifty-page action item list of action items, but hey what's another ten pages from these fine folks! The air was quite condescending.

After the meeting, I was pulled to the side and advised that the "loop team" had discovered that some standards were not being followed. They wanted me to fire one of my managers, and they would leave a guy back to cover until I found a replacement. We wrapped up and headed to the airport. More stuffy conversation on the way.

Everyone but my boss had gotten on the jet. With a smile on his face, my boss said to me, "I'm looking forward to seeing your progress." I said, "You know what? I'm looking forward to seeing that jet lift off the runway!" He laughed. I asked him, "What was that?" He said, "What do you mean?" I said, "What we just did?" He said, "It's a loop." I said, "OK, but I have a list of hundreds of things to fix. The plant is in a mess. We know that, because you hired me to fix it. Was it really necessary to fly a group of people here to point out what we already know, that the plant is doing terrible?" He said, "Don't worry about it. Just fix it. I got your back." A short time later, that president left the company.

Later on, a wise and experienced plant manager taught me how he handled the "loops." He said to say, "Thanks for pointing that out. What can you do to help me?" whenever the loop folks pointed out things, or "Could you stay at my plant for a few weeks and help me fix that?" I, in fact, started to do that from then on. Anytime someone would point out a problem, I immediately requested that they stay at the plant to help, or at least take responsibility for helping address the issue. This reduced the occurrence of people trying to earn points for pointing out problems.

This great small boat company that led the way in volume and cost control in the industry was seriously losing its way. During my career, I have noticed a certain dynamic that occurred on more than one occasion. That dynamic is when a company's management allows a factory, facility, or even a department or a function to deteriorate to a situation where serious operational issues develop that necessitate a turn-around. The dynamic I am describing lies in that the rest of the company turns against the poorly performing facility or department. They literally "pile on." People run up and down the hallways of corporate, proclaiming another instance where the poorly performing facility has missed a deadline, fallen short of a metric, has not met a schedule, or what have you. Now in some ways I understand that part of an organization not performing does drive inconveniences into other areas, so having frustrations arise is understandable. However, being a "turn-around guy," I can tell you that this dynamic goes directly to what the company wants, which is to turn around this operation to performance and/or profitability as fast as possible. There were times when I spent a couple of years sustaining political attacks over issues that I was there to fix, yet no attacks were directed at the source of the problems. I lived through the attacks and the people running up and down the hallways, but I never have quite understood this dynamic.

Also, when you are hired to turn around a mess or lead a transformation, management expertise comes out of the woodwork. People love to give advice and tell you how to turn it around, but this type of advice is much more plentiful than the will to roll up sleeves and fix things.

So here is my advice to any president who has a division or factory that is not performing and wants to get it fixed as fast as possible: First, take responsibility for the situation. Ask everyone to direct all attacks and criticisms of the poorly performing operation toward you. You are in charge, the buck stops with you, and this happened under your watch. You take the heat. Second, direct everyone in the organization to do all they can to assist the facility during the turn-around. Hold the facility accountable for improvement, but ask the facility how you can help. Focus the organization on doing everything it can to get this facility or function to operate effectively. This approach is not only the right thing to do, but it is also the best course of action. You simply cannot build up an organization and tear it down at the same time.

THE TRANSFORMATION BEGINS

Leading a Lean transformation while turning around a poorly performing factory is not for the meek. It is unbelievably stressful and challenging; yet it is one of the most rewarding things you could ever do. I had a big list of things to do, and I was a rookie at Lean Six Sigma. While staying in a hotel during the first few months, I would sit in a restaurant until 2:00 a.m., drinking coffee and writing down list upon list of things to do. Then I would wake up in the middle of the night in my hotel room and write down some more things to do. Then during the day, all day, I was on the plant floor leading people to get it done. Living in the Gemba. My mind was racing constantly. It was all good, however. Do this. Fix that. Make a plan. What can we do? It really was Plan, Do, Check, Act (PDCA) over and over and over again. Develop this. Remember to do that. I must have had a dozen legal pads and a stack of books.

When I first started, I would get a telephone call almost every day from my boss. I was usually in the bottom of a boat with boat builders or in the middle of some production problem, and I would get paged to take the call. Someone would say, "Hey Bill, you are being paged." I would get the page, find out it was my boss, make my way through the factory to my office, and take the call. He would ask, "How are your labor efficiencies?" "How's your materials variance?" "Where are you at on your schedule?" It was like a wife calling a fireman in the middle of a five-alarm fire to ask him what he wanted for dinner! After a few of these calls, I said to him, "Look, I can spend my day in the office scouring these reports and have accurate answers of bad numbers whenever you call, or I can go out into the plant and fix the things that are driving the numbers and give you an update once a week." I said, "If you want to ask what's wrong and what is being done to fix it, you can call me anytime and I will know, because I will be personally involved. I won't know what the 'numbers' are, but I do know that they will be bad, and they're going to be bad for a while." As he told me that he completely agreed with me and would have my back, my respect for him began to grow.

5S

The housekeeping in the factory was horrendous. You could find trash everywhere but in the dumpsters! There are very few challenges as great as instilling the value of housekeeping (or 5S; X) in a plant that has never done it. I made some copies of the pages with the 5S philosophy in Iwao Kobayashi's book *20 Keys to Workplace Improvement* and passed them out to the leadership team. We sorted and sorted. There were cabinets everywhere. They were stuffed with excess inventory, hoards of fasteners, parts, trash, tools, you name it. The guys did not want to get rid of them. We agreed to keep them as long as they were kept organized. Within a couple of weeks, we all realized that it was not worth the trouble and hauled them out with forklifts. They were large, and the team was surprised at the floor space we freed up. Seeing the plant open up more and more and the cleanliness and orderliness began to build pride in the team. This kind of thing is rewarding to any leader. Cleaning up the workplace always builds morale.

In the middle of the plant was some sort of terrace-like rack that was full of fasteners. There was box after box broken open, with other boxes of fasteners spilling into or onto other boxes' contents. There were fasteners everywhere within twenty feet of this area. At all times there were boat builders digging through them. Behind this fastener rack, there was a huge pile of fasteners in buckets, boxes, cans, tool trays, you name it, that had been removed from the cabinets we had removed from the plant.

Throughout the workday, the workers would throw trash and debris on the floor while working. Wrappers, cut-offs, wire tie snips, fasteners, cut-outs, boxes, and you name it, were everywhere by the end of each workday. Then for about forty-five minutes at the end of the shift, everyone would grab a broom, sweep, and pick up most of the debris. There were no trash-cans. There were a couple of dumpsters, and everyone walked an average of fifty feet to empty each dustpan.

We sorted and sorted, and the guys started to get into it. They began to challenge each other about who was keeping what and whether they really needed it. Team building was already beginning. I challenged them again and again as to whether enough items had been tagged. Each time, they went back at it and tagged more items. There were two large wooden boxes of air tools in front of maintenance. There was also a host of other tools there, caulking guns, saws, saw blades, crimpers, jigs, repair parts, you name it. Pallets and boxes of fiberglass were delivered back to the stock room. Excess and leftover components from previous boats were delivered by the dozen. I had to calm down the materials supervisor several times. All these items would have to be identified and returned to stock, and in many cases returned to the supplier. It was becoming obvious that the Bills of material (BOMs) and associated pick-lists for the product were not accurate. More on that later.

We had area meetings, encouraged the employees, and showed appreciation for their efforts. 5S is a lot of work while trying to meet production. However, most of the employees were reacting positively to the effort. We explained the benefits of an organized work area: how it was safer, and that they would spend less time looking for needed tools and materials. The yard behind the plant was piled high with items sorted out of the plant.

There was a center aisle in the plant from one end to the other, about 600 feet long. The team painted an eight-inch yellow line along each side of the aisle. We rounded up dumpsters and placed one about every forty-five feet up and down the aisle. They were all lined up on the yellow lines. It looked great. Trashcans were placed in each area, and workers were asked to throw all trash into the cans or dumpsters. We asked that nothing be thrown on the floor during the process. We soon discovered that the boat builders were throwing the same debris on the floors of boats during the process. So we placed cardboard boxes in each boat and asked that all debris be placed in them. Each boat builder was required to have his own personal trash box.

The dynamics of people in the work environment are fascinating. To me, people are people; and by and far, most of them are good and want to do a great job. Implementing 5S always uncovers human nature. When we put the trashcans in the work areas and asked employees to not throw anything on the floor, but instead put all debris right into the trashcans, we did receive some interesting comments and suggestions. "What does it matter? We sweep up at the end of the day anyways." We would explain that debris on the floor is a slip-and-trip hazard, and spending time sweeping up things at the end of the day was wasteful and unnecessary. It is just as easy to put something into a trashcan as it is to throw it on the floor. The employees also shared the stress of making the schedule, and I explained to them that freeing up three people for forty-five minutes from sweeping up trash that did not need to be there would allow us to get more work done and lessen the load of making the schedule. We explained that working cleaner and keeping areas clean continually throughout the workday was a much better approach. With a simple explanation, most employees understood and embraced the change, and those who did encouraged and coached their teammates who did not. Employees often do a much better job of convincing or teaching their teammates new concepts than leadership.

The maintenance room was in the middle of the plant and somewhat inconspicuously located. There was a tool window and when you looked inside, the view gave the impression that it was just a small room, maybe six by eighteen feet. However, there was a much larger room behind it. Being new to the plant, it was a while before I learned of the backroom. One day, I saw an employee go through a door under a stairwell and close it. Being curious, I decided to see what was behind that door. The door was locked. I knocked, but nothing happened. There was a glass window on the door, but it had been spray-painted over. I knocked again. Still nothing. I went to the key box in my office and saw a "maintenance" key, so I grabbed it.

The next day while walking in the plant, I approached the door and opened it with the key. To my surprise, there was a "maintenance room" with about six guys in it who looked pretty sheepish when I came in. We are talking about eyeballs the size of golf balls! "Hi Bill, how's it going?" "Good," I said. "So this is the maintenance room." There was a toaster oven, a microwave oven, and there was stuff everywhere. It was a perfect before-5S picture! I met with the leadership team and we assembled the maintenance team together and brought them up to where the rest of the plant was with 5S. They basically did a stand-alone 5S effort in the room. Boy, did they haul the stuff out of there! We also asked them to use the lunchroom for cooking, and to scrape the paint off of the window so people could see in. Change was in the air.

The plant was starting to take shape as the 5S effort progressed, and a lot of the employees were really getting into it. As I walked through the plant, employees would stop me, give me suggestions, and make positive comments about the direction. As always, there were some naysayers or grumblers, but that was to be expected and I budgeted my time coaching them and trying to bring them along.

As a leader, I lived in the Gemba. I like to think of myself as a positive, enthusiastic, engaging leader, and I am relentless in leadership. It is my passion. Continually reinforcing new values such as an organized workplace requires leadership, so I spent a lot of my time doing just that. I must admit that before first learning of 5S in Mr. Kobayashi's book, I was really big on housekeeping. We all know that 5S is much more than housekeeping, but the being clean and organized part of the 5S philosophy won me over quickly. If the 5S process was not being followed, you could be sure that I would notice. I do not know why, but I have an innate ability to notice what is not the way it should be is. As a leader, I had to manage this ability to notice things that are not the way they should be because it can quite easily go to only seeing what is wrong, and a couple of trusted leaders in those years thought enough of me to bring it to my attention so I could learn. Some of the guys took advantage of my ability to spot a gum wrapper at a hundred yards. They would sneak into other areas and put some trash or scrap on the floor and wait for me to come by. I would see the trash and call to begin a 5S talk with the team leader, and then would hear giggling from across the center aisle. Ahh, esprit de corps! Getting a team to have fun with change is a good thing. Fortunately, another thing that is innate in me is a sense of humor. I love to laugh, I look for the fun in everything, and I use humor in leadership.

I had been reinforcing 5S quite a bit over a few weeks. For a couple of weeks, we emptied out the dumpsters on the floor each day to see what we were throwing away. We found hundreds of dollars worth of plastic ties, rubber gloves, stainless steel fasteners, carpet knife blades, full or half-full caulking tubes, you name it. Now, as the leader, I always helped sweep up and put all the trash back into the dumpsters. It was fun and also a great chance to interact. It is a big deal when the plant manager is sweeping floors with the guys. We would tag some of the items with the cost indicated, and we set up a display in the lunchroom to educate the employees. They responded positively and, after a period of time, the leadership team would spot-check a randomly selected dumpster once a month. It has been said that if someone goes through your residence's trash for thirty days, they will know things about you that you do not want them to know. Believe me, go through any factory's dumpsters and you will learn a lot, and it will be very telling.

As the team progressed through 5S, we started receiving a lot of positive comments from visitors to the plant such as suppliers, people from the community, and even corporate. Seeing a plant before *and* after a 5S implementation can be a powerful thing. Although the plant was becoming amazingly clean and organized, we still had a long way to go, but this was an awesome start.

The plant manufactured sport yachts up to just under fifty feet in length. WIP was everywhere and was not a surprise as it was just another symptom of an out-of-control process. The primary bottleneck in the plant was final assembly. The lamination department laminated the hulls, decks, and small fiberglass parts. They could laminate bad parts more quickly than final assembly could assemble them, and, by the way, departments earned incentive by department, and quality was not part of the incentive pay system. In effect, the company paid lamination to put out bad parts. What a mess!

Although no one ever made incentive, they died trying. As a result of all of this, there were quite a number of decks and hulls sitting outside the plant awaiting final assembly. These hulls and decks would fill with water, sometimes it would freeze, and on and on. Remember the waste of over-production? Well, this was it. On the back of some of the larger decks were built-in storage tubs; I would guess they held about fifty gallons of storage for the customer. The drain holes had not been drilled into these deck tubs, so they would fill with water. Thus, when final assembly needed a deck, they would dispatch five or six guys to the parking lot to bring one in. One day, I was in the Gemba and saw the big garage door open up and five or six guys came into the plant pushing a deck (it was on a dolly). The tubs were full of water, and down the center aisle these guys went with the deck, water and ice sloshing out the whole way! I said, "Guys, what are we doing?" "We're bringing in a deck, Bill." "I can see that. Why don't we drill the deck drain holes before we bring it into the plant?" I asked. "I guess we could do that." I requested that we do that from then on. There wasn't any standard work to tell the hole-cutters to cut those holes in the hole-cutting room. They were only doing what they were taught or asked to do. We had a factory where nearly 400 people were doing things they thought should be done, or not be done, or between the way they thought they should be done.

The plant progressed quite nicely through the 5S process. It did take three or four months of relentless positive reinforcement, but the plant looked great, was much more organized, and was becoming more efficient as a result. The plant started receiving positive comments from visiting suppliers and dealers on the cleanliness of the plant. That was great for the employees. The plant had taken a lot of criticism over the years. All they needed was leadership. The team repainted the yellow stripes anytime we had visitors to the plant. One thing that bothered me was that you could not tell where the front of the plant was. If you drove completely around the plant, you would not know what door was the front door. I requested to buy a sign for the front of the plant but it was denied. I challenged the guys to make a sign. They laminated two panels of white fiberglass about three by eight feet each, welded a frame, and Gordy, one of our manufacturing engineers, painted the company name and logo on it. We put it out front, and everyone took pictures in front of it. It meant a lot. I remember someone mentioning that the company had a flagpole fund, so I ordered a nice flagpole. It actually cost about $3,500. The maintenance team poured a small slab of concrete, mounted the pole, and even landscaped a small

FIGURE 9.1
The Maxum Yachts plant in Salisbury, MD.

flowerbed around it. I told the guys to get the flag so we could put it up. They said no, and asked me to meet them by the pole the next morning at 6:00 a.m. So, I did. To my surprise, some of the guys were in the National Guard and Reserves. There they were in their dress uniforms and they presented the flag and dedicated it in a really special, short ceremony. The plant sign and the flagpole meant a lot to the employees (see Figure 9.1). I could really see the pride growing in the plant.

We celebrated the 5S implementation with a pig roast. The maintenance guys built a roasting pit with concrete blocks and stayed up all night roasting the pig. The rest of the leadership team and I served the employees. This was a fun event, and you could tell that the employees appreciated it. I moved from table to table, visiting with the employees. I always enjoyed getting to know them. Just as important is for employees to know their leader. If they know you well, and respect you, they will go all-out for you and the organization. Lots of fun.

QUALITY: STABILIZING THE PROCESSES

The quality team continued to "write up boats." When the supervisor said the boat was ready, the quality auditors would get out their flashlights and clipboards and scour every inch of the boat, writing down every defect they could find and identifying many of them with a China marker or tag. This only happened at the end of the process. The QA team did not go anywhere else in the process. The lists would often be 300 or more defects. Once the list was complete, QA would give it to the supervisor, and he or she would assign boat builders to address the defects. When they were all worked off, the QA team would "buy them off," which meant to confirm that the defects were, in fact, addressed. To this day, I have never liked the term "buy off." After the "buy off," QA would give a list of defects that were not addressed to their satisfaction, and also any new defects they identified that they had missed previously. This process went through several iterations before the boat was finally completed.

Now mind you, this process was done in the absence of clearly defined standards. There were some manufacturing standards in the company, but they were not maintained, nor were they complete. A large part of the QA process was dependent upon what the individual auditors thought "looked right." It was crazy. Often, there would be disagreements between the auditors and supervisors that had to be settled by the QA manager; and if they could not come to an agreed-upon conclusion, they would come to me and ask me to make a ruling. I will say that I am obsessive about product quality, and I was fortunate to have a good guy for a quality manager and we almost always reached an amicable solution. However, the absence of a "formal quality process and standards" generated a lot of wasted discussions that would never have occurred with a process and standards in place.

This process, or no process other than an inspection on the end of the line, was not working. It also hurt morale. Also endemic to this process was a total lack of trust. The QA team was considered part of corporate and not part of the plant team. List upon list, and more and more tension. On one occasion, a frustrated production manager brought me onto a fifty-foot boat and walked me to the head unit (bathroom), which was a room molded out of fiberglass. The manager pointed out to me a small scratch that you had to use a bright light to see down low behind and under the toilet where you had to be limber and hold your neck just right to find. First, the boat's owner would never see that scratch. There would never be bright light in this area; and to get a buffer or sander to this scratch would probably result in creating four more scratches. I could think of no conceivable situation where the owner of this boat would be in a physical position to see this imperfection. Of course, there was no standard. Just a wonder of the QA auditor as to whether or not it should be reworked. One standard I always applied was to ask myself the following: If I owned that boat, would I want it fixed? I absolutely would not want a bunch of air hoses dragged through my $400,000 boat to repair a small scratch that I probably could not find, or would never see.

As I worked with Joe, our quality manager, we began to build up a lot of trust. I really had to put my money where my mouth was with product quality. Actually, it was easy because I have a very high personal standard for quality. To me, quality is integrity. I have never knowingly passed on a defect in my career. I have made some tough quality calls, usually where the fix was worse than the condition, but never allowed a knowingly defective product to be shipped. Once Joe really knew that we had the exact same goal in mind, we began a very close and enjoyable working relationship. His product knowledge was incredible and, combined with my familiarity with quality systems, we made a great team.

Another facet of this quality situation was that during the rework of these hundreds of defects, air hoses, toolboxes, equipment, and who knows what were brought on and into the boats, often creating more damage in addition to what was being reworked. The boats did ultimately look good before they left, but I felt they were already "old" while they were actually new. This did not sit well with me because *my* name was on this product.

It was obvious the facility had to have a defined quality process in order to grow. Joe and I went into the conference room day after day and hammered out a quality process. We got out everything we could find on product standards, specifications, and anything related to quality. It turned out that we found some very useful information. We found a great product and process standards manual, but it was outdated; and although a lot no longer applied or was missing, there was a lot that did apply.

Building on what I learned at Outboard Marine Corporation about quality and the use of critical-to-quality characteristics on drawings, I suggested that we develop a list of all product and process characteristics that if not correct to a standard, or within a tolerance, would cause one of five situations:

1. Overall product or component failure
2. A significant downstream production interruption
3. A safety hazard
4. Significant customer dissatisfaction
5. Violation of a regulation or governing body requirement

We developed a list of critical product characteristics for process points throughout the process. Critical product characteristics were checked per part, critical process characteristics were checked per time period (such as "x" times per day or week), etc. We then obtained or created a standard for each check. If there was a current company standard, we used it. If there was one but it was outdated, we updated it; and if there was no standard at all, we created one.

Next, we put quality documents on the back of each boat at the beginning of the process. The employees were required to complete the quality document for their station and notify QA. Then QA would double-check. This way, anything critical to the product or process was checked twice, and the QA documents with the standards on them informed the employees what was expected. The employees embraced the new process. Everyone did.

Once all audit forms were in place and in use, we recorded how many critical characteristics were correct the first time. We measured by area and by boat. We called it "First Time Pass" (sorry about the slogan, Dr. Deming), and the intent was to measure how many things we got right the first time. The scores were posted on dashboards where everyone could see them and they were reviewed as a team daily. We also added in 5S process checks to add to our 5S sustainment efforts.

QUARTER DRILLS: PRODUCTION STATUS BOARDS

The plant was gaining some momentum with the 5S implementation and First Time Pass disciplines in place. There was a long way to go to fully mature these efforts, but they were getting stronger by the day. We were also approaching being ready to use data to use to help us track our progress and tell us where to focus our improvement activities. We had the FTP (First Time Pass) quality data, labor efficiency data, safety data, and plant financial statements.

We created production status boards and held what we called "quarter drills" four times per shift. (See Figure 9.2.) The drills or "huddles" were stand-up and quick, usually about five minutes, and conducted in front of a production status board on which the areas FTP, Labor efficiency, Customer feedback, and any issues affecting the area were displayed. It took the teams a while to get the hang of conducting the drills with ease. During the first drill, the team reviewed the workplan for the day. Every two hours after that, they met in front of their board and discussed the progress for the daily workplan. Any issues or problems were raised and

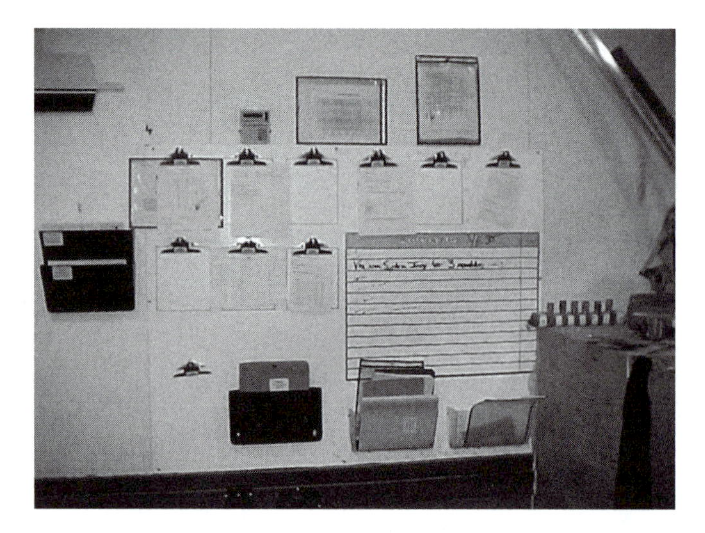

FIGURE 9.2
Example of a production board.

decisions to shift or acquire more resources were made. The team leader would coordinate resolving the issues and keep the progress going. Some of the employees liked them, while others did not. The drills raised the level of healthy accountability, and thus raised the bar for the team and the factory as a whole. It was good.

MANUFACTURING ENGINEERS/PROCESS IMPROVEMENT TECHNICIANS

Now that the 5S and basic quality processes were underway, we needed a resource plan to find the root causes of low first time pass (FTP) scores, whether for product or process, and lead improvement efforts to address them. We created a manufacturing engineering/process improvement team. We did this by selecting highly talented boat builders from each area who were good problem solvers, had good people skills, and were respected by employees. We assigned manufacturing engineer process improvement technicians to lamination, assembly, cabinet/upholstery, and electrical hook-ups. We also assigned MEs to maintain standards and documentation. We trained them in cause and effect, brainstorming, fail-safing/poka-yoke, PDCA, and quality standards.

Each week we would review the FTP scores from the quality process, and the team would make decisions on where to assign the ME team members to assist operators and team leaders in determining root causes and appropriate solutions. This is where we implemented the plan, do, check, adjust (PDCA) concept from Dr. Deming. The MEs also assisted in creating jigs and fixtures to fail-safe certain process operations. They determined what equipment or tools were needed to address product issues and trained operators on how to use it.

LEAN TRAINING

During my time at Chris-Craft, I had obtained a small pocket guide called *Waste-Chasers*. I still have it to this day. It was a quick reference on how to do PDCA, Pareto, Affinity, waste identification, run diagram, scatterplots, histograms, you name it. It was great. I decided to contact the company that created it, the Conway Company. The Conway Company was founded by Bill Conway. He had worked personally with Dr. Deming and clearly understood Lean principles.

The Conway Company conducted training regularly. I arranged for Joe and to go to three days of training on measuring and quantifying waste, process improvement using Lean tools, and managing people in a Lean culture. We got a package of three training seats. I offered the extra seat to a fellow plant manager, and he decided to come with us. In addition, a corporate HR representative decided to tag along. The training was in Coconut Grove, Florida, and it was great. The concept of understanding waste was fascinating, along with learning how to identify it. We learned how to more effectively measure processes, identify waste, and create Pareto analysis, brainstorm, and lead improvement activities. This was great stuff. We were really fired up about getting back to the plant and putting these tools into action.

We recognized the importance of training and involving all employees in improvement activities, so we ordered a twelve-week training course with videos and slides on basic process improvement skills with a focus on eliminating waste. The program consisted of weekly three-hour training sessions and covered all the tools we studied in the off-site training. We put all supervisors, team leaders, and our MEs through the training.

Joe and I teamed up to teach and facilitate these training sessions. We trained all our supervisors, crew leaders, and some employees. We then started a "War on Waste," a term taken from a book written by Bill Conway who started Conway Management, a consulting company specializing in taking waste out of organizations. He had known and worked personally with Dr. Edwards Deming. Joe and I had eaten lunch with Mr. Conway at his seminar, and he talked about knowing Dr. Deming.

We trained employees in identifying waste and chartered improvement projects to reduce wastes and costs. They developed ways to fail-safe or mistake-proof operations, reduce process times by eliminating unnecessary steps, and reduce process times and improve quality by creating unique jigs and fixtures. These ideas came from the minds of our employees; and let me tell you, there were some great ideas.

Boot Stripe Rework Reduction: Defects, Non-Value-Added Rework

We created employee teams and launched projects in the lowest FTP parts of each of the areas in the plant. What was awesome was that the team leaders or supervisors led the projects. They used the tools of brainstorming and root cause analysis using fishbone diagrams. (See Figure 9.3.) One supervisor nicknamed a project "boot stripe brainstorming," which had

FIGURE 9.3
Example of a fishbone diagram completed by employees.

to do with eliminating rework on the feature stripe on the lower sides of boats. The process for putting the stripe on the boat involved taping off the stripe shape in the mold, spraying gelcoat, pulling the tape off not too soon or you got smudges, and not too late or you got tears because the gel had begun to catalyze. The type of tape used (glue residue) and touch-up procedures, along with some other variables, came into play.

This effort produced numerous awesome process improvements that improved quality, and reduced rework and unnecessary processing. Again, the thing that fascinated me the most was that these solutions were coming from *our employees*. We also applied 5S and developed numerous jigs and fixtures to reduce fit issues and rework.

Lesson Learned

- Never underestimate the incredible solutions that your employees will come up with after relatively minimal training in basic problem-solving approaches.
- Cycle time reduction: inventory, non-value-added, defects.

WHERE TO START

The plant produced a forty-six-foot sport yacht that was plagued with issues. There was a large backlog of sales, and these boats retailed for nearly half a million dollars each. So there was some revenue pressure present. The plant had about ten of these boats in process at any given time. The cycle time of these boats was over six months. For the most part, there were seven boats in assembly alone. That's seven $400K+ boats in WIP. They were all in station with a different crew inside each one. Sometimes crews moved; sometimes they didn't. They were doing an immense amount of work inside a boat. Imagine going into a small bathroom in your house with a toolbox and a pile of wood and building a set of cabinets in there. Get the picture? The boats were full of toolboxes, air hoses, boxes, materials, and/or sawhorses. There was an inordinate amount of damage being done to them. There were people standing on things to put the vinyl ceilings in the boat or fit pads to the walls. Glue was being sprayed, and padded vinyl was being laminated to raw fiberglass walls; and if there was a bump, there was some grinding going on. People were installing refrigerators, stoves, microwaves, stereos, speakers, you name it. There was little accurate information or documentation available. There were usually a couple of people outside the boat working on something on the floor trying to get it ready to install on the boat. There were big, plastic engine shipping crates stuffed with thousands of dollars of components and parts piled on top of each other. There was always someone inside rooting around trying to find something. If they did not find it, they went to the warehouse to get the part they needed. Then when the box empties, that part would be there in the box and have to be handled and returned to the warehouse. There were people putting gages in dash panels on the top of the boat. The person managing the BOM for the boat was 3,000 miles away. Are you getting the picture yet? Other than that, things were actually going quite well!

So where do I start? We knew we had to shorten the cycle time of these boats and reduce the WIP, or we would never win. The WIP footprint was sucking the oxygen out of the plant. So we decided to focus on four things:

1. Move all work possible out of the boats.
2. Look at all the things that are done to things right after they are taken onto the boat for installation and move that out of the boat. Look at everything that is done to things right before it goes onto the boat and move it to the sub-assembly or fabrication department. Get it out of the critical path.
3. Cover everything we can to protect it. Give the boat builders a padded tray to lay tools on instead of on upholstery, countertops, or furniture.
4. Move all work as far back in the process as possible.

There was one problem with this plan. We did not have a sub-assembly department. That was a problem. We needed one. There was an area in the middle of the plant just off the assembly area that had a CNC operation in it. There was another adjacent area that was actually closer to the point of use for much of the material that was cut on it. So we set up a Kaizen event to create a fiberglass sub-assembly area. We set up padded benches and work areas to build dash units and easily install gages, switches, and controls. We started to build head units, complete with padding and all. We also used pressurized air from a tank to test plumbing for leaks prior to installation. Finding and repairing leaks on a finished forty-six-foot boat—or any boat for that matter—is time consuming and less than fun. We set up a 12V battery and radio to test the speakers and lights on radar wings prior to installation. This department began to install hinges and hardware on sub-assemblies before they went to assembly. All this work was moved outside the boat.

During this time, we also developed a manufacturing engineering (ME) team. We selected some of the most talented boat builders we had from each area—lamination, cabinet building, assembly, and electrical. These guys were good and a pleasure to work with. They could do anything. In fact, as a team, we nearly redesigned the boats. We had no choice.

Next, we developed modules for anything and everything we could—yet still sticking to our strategy of moving everything we could out of the boat and actually the assembly area itself. We called our strategy in assembly "pick and stick." The only thing we wanted to do was pick up stuff and stick it in the boat, accomplish a couple of hook-ups, and complete the installation.

Our fabrication department cut and assembled cabinets, galleys, and entertainment centers. We asked them to install the microwaves, refrigerators, stoves, sinks, faucets, stereos, televisions, and anything else that went into these units when they built it. We went from installing ten things on the boat when the entertainment center was installed to one thing.

Quick story. While all this was going on, I was trying to instill pride in workmanship in our boat builders. They were great people, but they only knew what they were taught. One day while on a boat, I noticed quite a morass of wires and harnesses along the wall. There were wire ties everywhere and screws all over the place. So I called the team together. I said something like, "Hey guys, let's think how would you want that wiring to look if it was your boat." One guy immediately said, "I wouldn't worry about it. You can't see it." A couple of his teammates jumped in and said, "No way! If that was my boat, I would want it to look like a racecar under the hood. It all needs to look good." I asked them to think about that, and I knew they would do the right thing. A few weeks later, I was walking past a boat and heard a voice call out, "Hey Bill, come here a minute." I went on the boat and a boat builder pointed to the side of the boat and asked, "How's that for boat building, Bill?" All the wires were straight; the wire ties were equidistant apart; the screws were spaced equally apart. It looked great! "Guys, that is awesome!" I said. "Great job. I am proud to be on the same team." When I got off of the boat, something caught my eye. All those screws equally placed inside the boat were sticking through the side of the boat!

As we reduced the WIP from implementing Lean, we freed up some people. We made them "water spiders" and assigned them the responsibility of doing whatever they could to make it so that operators did not have to leave their workstations. Whether it was going to the stock room to obtain a needed part, coordinating getting the BOM or kit list corrected, or getting a tool from the tool room, water spiders did it. It kind of "turbo-charged" our efforts. I have practiced the approach of converting the first people you free up from Lean activities to process improvement support roles over my entire career. It is always fun when you get to this point. It really does give your improvement activities a "shot in the arm."

We had "modulized" everything in every boat we could, with particular focus on the forty-six footer. We had taken a ton of work out of the assembly lines and moved it back in the process or out of the process. We were not prepared to build this boat in a much shorter cycle time. Ultimately, we achieved three weeks, but we started moving to four weeks. Well, in order to go from seven boats or WIP stations in the assembly department, we had to complete three of them while trying to keep to the Takt time. We had to create the modules for these boats, the cabinetry, and all the sub-assemblies, along with the normal production schedule. We also were not all that efficient at building these modules and sub-assemblies yet. We were training new people in new jobs, and it was tough. But, in order to move forward, we needed to do this. So we worked a lot of overtime and spent a ton of labor to get to the new WIP footprint.

Well, all that overtime and labor gave us a huge labor variance for the month. It was something like $100,000 or more. A few days later, I was paged to my office for a phone call. It was my boss. He said, "The president is pissed off about your plant's labor variance, and he wants an explanation." I said, "OK, I'll get something together and send it to him." He said, "No need to do that. We are on our way to the corporate jet and will be in your plant in about four hours. You can tell him yourself, and you better have good explanation."

Oh, the adrenaline of the most competitive game in the world—manufacturing! I called my team together and said, "Guys, the president is pissed about our labor variance, and he will be here in about three hours on the jet." I told them they might be getting a new plant manager. I fired them up and said, "Let's show him who we are and what we have done."

The whole plant was sweeping. We had a clean plant anyway, but these guys put a shine on it. The employees left at the end of the shift, leaving the leadership team to wait for the executive staff. I went to the airport; it was about twenty minutes from the plant. I picked them up at the airport. My boss winked at me and grinned. I greeted the president; he was not smiling. I tried to make small talk on the way to the plant, but all answers were short.

When we got to the plant's conference room, the president said, "Bill, your costs are out of control. Your labor is atrocious. This must be fixed." I said, "I agree, and the reason our labor was so high this month is because we are fixing it. John, if you don't mind, I would like to walk through the plant with you and let the guys show you what is going on." He agreed.

I walked him out into the plant and went directly to lamination. Now, I am not bragging—well, yes I am—but I can guarantee you that he was now standing in the cleanest and most organized lamination department he had ever been in. I immediately introduced him to Tony, Brian, and Jeff, the team leaders, and they each took him through their whiteboards, reviewed the process improvements they were making in their areas, as well as their progress on quality. These guys were doing awesome. I was watching the president, and I could tell he was getting into this.

We showed him the sub-assembly department and explained our strategy. We took him to fabrication and let the team leaders take him through just as the lamination guys had. And then we went to the assembly department. It was clean and organized, and we were almost down to our four WIP boats. We stood in the middle aisle in assembly, and I explained our effort to reduce cycle time and how we had to work ahead to burn the WIP down to a four-boat footprint, and that was why we had spent all those labor dollars. He asked me, "Did you have to do this all in one month?" I explained that we could not have let this situation continue. It would cost us much more money the longer we took to fix these issues. Our president stood in the center aisle of our plant and said, "This place fires me up! This is awesome!" You should have seen the look on my guys' faces. They were so proud. What a great feeling. It is hard to describe how fun it really was.

The president of the company was so impressed that he directed a companywide "War on Waste." The entire company was adopting the strategy of *our* plant and *our* team. I was so proud for the team. The pride in the team and employees continued to grow, and it was so rewarding to hear people talk about the plant in such positive ways. It was even spreading through the local community. The local ASQ (American Society for Quality) chapter came and met at a different company each month. The meeting involved some refreshments, a facility tour, and a discussion of chapter business and events. Everyone so much enjoyed touring the plant and seeing the boats, that we were requested to have the monthly meeting at the plant at least three times. Good fun!

Shortly after that, I actually received a note from one of the managers at Pier 33. She scribbled, "P.S. The boats are looking better," on the back of a boat part. That note was the first positive comment or gesture from that dealer. It was hard earned, and it felt good to get it. A few weeks later, I was getting ready to leave for the evening when I got a call from a dealer. Ending the day with a call from a dealer was not much fun during a turnaround. He said that he had just taken delivery of one of our boats and had some questions that he wanted answered. I grabbed my tablet and prepared to start listing his issues. I knew the routine. I told him I had a tablet and was ready to start writing. The dealer said, "What are you doing up there?" I said, "What do you mean?" He said, "I got a boat in today from your plant, and I couldn't find anything wrong with it!" Gosh, did that feel good! The amount of work it took to get that comment was unbelievable. Although we had a long way to go, the tide was turning. I could not wait to tell the team about that telephone call.

One waste that commonly occurs is in having excess amounts of tools. To an experienced production boat builder, and especially a plant manager, the sounds in a plant say something. There are two sounds that you should hardly ever hear in a final assembly operation: (1) the sound of a hole saw cutting a hole or the sound of a saber saw. These operations are supposed to be done earlier in the process in the hole-cutting room. I call this "the symphony of saws." There are a myriad of holes drilled and cut into boats during the production process. Many assembly operations involve cutting a hole or holes into something. Let's say there are seventy-five people in an assembly department. In the absence of a hole-cutting work center, that means seventy-five drills and hole-saw blade sets must be purchased and supplied to boat builders, and they may mar finishes when laid on the boat. Boat builders will also spend time changing out hole-saw blades. On the other hand, if the holes are cut in a hole-cutting work center, only three or four sets of drills with hole saws will be needed. There will be one drill per hole-saw size too, so it will not be necessary to spend time changing out blades for different-sized holes. Talk about set-up reduction. When I first arrived at the plant, I could hear eight or nine holes saws going at once at all times. We solved that by setting up a hole cutting area and moving it all back to that area. If anyone heard a hole saw in the assembly lines, they went right to it and found out what the problem was. It became an "audio" factory. The plugs that come from holes also create a slipping hazard on the production floor. Slips, trips, and falls are the most common injury in production boat building.

The employees were really getting into the improvement activities. It seemed that every day as I was moving throughout the plant, I would hear, "Hey Bill," from some employee wanting to tell me about an idea that he or she had for an improvement or opportunity to eliminate waste. And they had some great ideas. Some of the boats had fiberglass liners. They were gel-coated and laminated like all other laminated parts. Then one day an employee said, "Hey Bill, if you want to see waste, why do we gelcoat all these liners and then just cover it all up with carpet?" He was onto something. Gelcoat cost close to $2 per pound, and there was the labor and equipment to set it up, apply it, and clean later. We initiated a project and stopped spraying gelcoat on liners.

I continued to study Just in Time, Lean, and quality. I was reading a book titled *Workplace Management* by Ishiro Kyaboshi, and it read something like this: of all the areas in production to manage, there is none more important than that of WIP…It must be minimized. You will benefit. A short time later, I was standing on the mezzanine with my production manager, Ronny Brown. We were stressed because we could not get parts moving through the plant. We had little flow. We were looking across the department at what seemed like a sea of hulls and decks in lamination. I said, "Ronny, it seems like everywhere I look, people are working on parts and more parts, but yet we can't get a part out of this department for nothing." He agreed. I told him about the book I was reading and that you had to trust that if you reduced your WIP, that things would speed up. Cutting your WIP requires a leap of faith. We had so much downtime, and so many parts that could not be worked on because of some defect, missing material, incorrect material, or damage. Having a lot of things to work on and keep people busy gave us some comfort. The natural instinct is to believe that as long as you are working on something, you are making progress. Not. But the feeling was real. On the other hand, what we were doing just was not working.

I asked him if he would be willing to try it and pointed out the progress we made with the forty-six footer. He was concerned but he agreed. We decided to put one part in the mold, one in the grinding and hole cutting room, and one in the Finish area for each model. Effectively, a fifty-foot mold was a great Kanban. You could not laminate another boat until the mold was empty. We would not pull any part from a mold unless a completed one left the department. We gobbled up our WIP and took a huge step forward. We were working on less and getting more done. It goes against one's instincts to believe that the less you work on at a time, the more you get done. However, nothing could be more true—unless you are my son working on homework.

During all of this, there was great pressure from corporate. The status of parts was entered into the company MRP system. Any time a boat did not complete an operation to schedule, we got a call. All they (corporate) cared about was numbers, at least the number of completions. It was stressful. As a leader, I was on the floor trying to teach people that if it is not done, if it is not right—while corporate only wanted the boats done.

GRINDING AND FLIPPING FIBERGLASS DECKS: WASTE, REWORK, TRANSPORTATION, MOTION

Rework was a huge issue in the plant for a variety of reasons. Some of the drivers were the process itself. Decks and liners, which are big fiberglass parts, were laminated in a mold, popped out of the mold, lifted with chains and a crane, and placed onto a dolly. They then were sent into the hole-cutting and grinding work center to have the edges ground down, trimmed, and rough glass ground down. It would be masked off with plastic and have the base coat applied to prevent anyone seeing raw glass on the boat. The masking process was labor intensive, and controlling overspray was a challenge. When overspray got on other parts, it required buffing and sometimes sanding to remove it. This was pure rework waste. Once grind and trim was complete, the deck was brought back out into lamination, flipped back over, and placed back onto the same cart. The cart had adjustable pads to hold the deck upright or upside down. Operators would readjust the pads for upright placement.

OK, so why can we not just base-coat the parts in the mold? Because they must be smooth. So why are they not smooth? Because the glass is rough. Why is it rough? It just is. Why can it not be smooth? It probably could. Is there any way it could be laminated more smoothly? Maybe... if the laminators paid more attention to detail during the fiberglass roll-out process. Why don't they? They have never really been asked. Can we eliminate all rough glass with more attention to detail? Not all, but probably most of it.

Why can we not grind the parts in the mold before so we can base-coat them in the mold and not have to move them into the hole-cutting area to do it? It will be way too much grinding dust in the middle of the department. If we focused on rolling glass more smoothly during lamination and got the parts to where they only required minor touch-up grinding, would we be able to do that? Maybe. Why don't we try? OK, we will.

We challenged the laminators to do a better job of rolling out the fiberglass and they did. We then began to just do minor touch-up grinding before the part was pulled from the mold. We assigned the laminators the responsibility to do the touch-up grinding. It is much easier and more pleasant to roll out smooth glass than it is to grind down rough glass, so there was a big incentive for the laminators to produce smoother roll-outs. The parts were now base-coated in the mold, which eliminated the waste of transporting the part into the hole-cutting room, and the overspray problem was immediately eliminated because the susceptible surfaces were not exposed when the part or parts were in the mold. Now somebody say, "win-win."

Now that we had the moving into and out of the hole-cutting/grind room and overspray issues addressed, we still had a big problem in reworking and repairing all the cracking, scuffs, and other damage to the parts incurred during the flipping process. The parts were moved into an area where they often had braces affixed to them for stiffening and then they were hooked with chain hoists and slowly lifted up until they were completely upright and then eased back down onto the floor. They were then lifted up and placed onto the cart upright.

Flipping a deck and setting it on the dolly took a couple of people about an hour to do. The operators would use chain hoist hooks to turn over the decks. During the flipping process, no matter how much care was taken, the decks would bend, and stress cracking, scratching, and scuffing would occur. This damage required labor-intensive repair later in the process. Additionally, flipping large parts was a safety hazard, even with the right equipment. So, unnecessary labor was being spent flipping and loading each deck twice onto a cart, and considerable rework was being generated due to cracking.

One day during a walk around the plant property, I observed some type of large steel frame. It was about sixty feet long by fifteen feet or wider, and about eight feet high. I kept trying to figure out what it was. Then one day it hit me. It was a fixture to flip the forty- to fifty-foot decks. It was huge. The problem was that it could not practically fit into the building, and the weight of it was outside the building's structural capacity so it could not be hung from hoists it required to operate. Most likely, this piece of equipment cost more than $10,000 to construct plus shipping and unloading costs, and finally the cost for disposal. All that money spent for a piece of equipment that was not even needed.

I challenged the team to "get out of the box" on how we could fix this problem. We brainstormed and discussed for a period of weeks. Once again, we began to ask, "Why?" Why do we flip decks? Because they must be upright, and they are upside down when they come out of the mold.

Is there anyway we can drop them upright from the mold? No. Why? You cannot get the deck dolly under the mold. Why? Because there is a spreader bar on the mold that is in the way. Could we remove one spreader bar to get the dolly under the part? Yes, if we affixed the mold frame to floor hooks. Can we do this? Not really. Why? The part deck will fall out when the mold is turned upside down. Can we stop that? We could try putting straps around the mold and part. We did, they did, and we were no longer in the deck flipping business.

The same type of process was used to flip liners, but instead, a turning mechanism with straps was used and the same types of damage were occurring. By just turning the part and mold over and dropping it out upside right negated the need to "flip" them. This project saved quite a bit of money in labor and reduced rework, and improved quality. It was Lean from the team. (See Table 9.2.)

FIBERGLASS SMALL PARTS

Producing small fiberglass parts was kicking in us in the rear end. Once again, we had parts everywhere but for the life of us, could not produce acceptable parts to schedule. Sport yachts are made up of a hull, deck, liner, and many small fiberglass parts. A rule of thumb was "one small part for every foot in length of the boat." A forty-six-foot boat, in fact, had about forty-six small parts. The plant demand was in hundreds of these parts per month. Small parts are complex, with a lot of geometry, and are really susceptible to air and other defects in the laminating process. In production boat building, small parts are a "big" deal. We could not schedule them effectively, and they constantly stalled downstream production operations. It was a huge problem. We had hundreds of parts in process and they were full of defects. The small parts laminators were producing parts, and lots of them. But they were so full of defects that the repair people could not fix them as fast as they could create them. Every day, the laminators would come in and produce a bunch of bad parts and then go home. We were working overtime, bringing in people from other departments. Tons of work, yet we were getting nothing done. Downstream departments had nothing to do and were constantly waiting on parts. This was a tough situation.

TABLE 9.2

Overview of Value Added, Non-Value Added, and Required Non-Value Added in the Deck Flipping Process

				Transport	Motion
Current State		**Cycle**			
Pop from mold and place on	RNVA	60	motion	60	
Move to hole cut/grind room	RNVA	10	transport	10	
Grind off rough spots	NVA	60	motion		60
Mask part off to prevent overspray	NVA	30	motion		30
Base-coat the part	VA	45			45
Remove the masking	NVA	10	motion		10
Move part to flipping area	RNVA	10	transport	10	
Flip part with flipping device	NVA	45	motion		45
Adjust cart, place part on cart	NVA	20	motion		20
Move back into hole cut/grind	NVA	10	transport	10	
Cut required holes	VA	90			90
Move to pre-finish	NVA	10	transport	10	
Repair stress cracks	NVA	720	motion		720
Remove overspray	NVA	60	motion		60
		1,180		100	1,080
Future State		**Cycle**			
Light grind rough spots in mold	NVA	20	motion		20
Base-coat part in mold	VA	45			45
Pop part, leave in mold	RNVA	30	motion		30
Remove spreader bar, spin	NVA	30	motion		30
Move to hole cut/grind room	RNVA	10	transport	10	
Cut required holes	VA	90			90
Move to pre-finish	NVA	10	motion		10
		235		10	225
Total Waste Eliminated		945	minutes		
Transport		90	minutes		
Motion		855	minutes		

Shortly after I attended some Lean training, I came back all fired up. We had done the one-piece flow exercise, and it really reinforced and highlighted the problems associated with excess WIP. My engineering manager and one of his team members had gone with me. We walked through the small parts area in the plant with them and observed the chaos. I suggested that we apply some of the training we just attended.

We outlined the process on a whiteboard, developed the WIP footprint, and set up a process. The process consisted of gel-coating/lamination, parts pulling, grinding and hole cutting, and finish detail. We decided to fill every mold we had with a part; after all, a mold is perfect Kanban. Once we reached our WIP footprint, we only pull a part if one has left the department. We had the plant scheduler number the parts, starting with the parts that were the most past due, and number them all the way to the least past due. Each day, he would add numbers to any new parts that may have been laminated.

Next, we set the WIP footprint at four parts in the grind and hole-cut room. We painted four yellow squares on the floor in the booth. Nothing could come in until something left. Square open, square filled. We also reemphasized 5S in this area. We worked with the operators to make their area highly organized. We developed overlays and fixtures to mistake-proof cutting holes. We also set up one drill per hole size so the operator did not have to constantly change out different-sized hole saws. There were five drills, each with a different-sized hole-cut saw bit in each, hanging on the wall within reach of the operator. We had the scheduler highlight important information on the sales orders and post them in the grinding booth so the operator could quickly identify what holes should be cut per the options on the boat. In the small parts finish department, we then set up six squares for parts. Parts could only be on a square or on a workbench. Again, nothing came into a square until something left. We placed one part on each of the eight finishers' workbenches. So there you have it: 100+ molds filled with parts ready to pull, four parts in the hole cut and grind room on their squares, six parts on the Kanban squares in the finish area, and one part per finisher's workbench.

Each time a part was delivered from the finish area, the finisher obtained a part from one of their squares. The hole cut/grind operator assisted in placing the part on the table. Once that happened, he would replace that part on the square with a ground part with holes cut from his area. When this happened, he immediately pulled the next part due from the mold, ground it, cut the appropriate holes, and placed it on the open Kanban square. By the end of the dayshift, and several iterations of this process, there were several open molds. The gelcoaters on the second and third shifts, upon seeing the open molds, would prep them and apply gelcoat to any and all open molds. The mold preppers would also check each part for necessary maintenance before gelcoating it. If it was due for maintenance, they would relocate it to the mold care area for maintenance.

The next day, the small parts laminators would laminate all gelcoated molds. And on and on. The beauty of Kanban to control WIP was realized. Something so simple is so powerful. For example, if the small parts laminators produce poor quality, fewer squares in the finish department will open up, less parts will be ground and cut, and there will be less open molds the next day. So they laminate the open molds for a couple of hours and spend the rest of their time learning finishing and assisting in correcting their defects. They will then come in early the next day or as required to catch back up on their lamination. This is the same for the grinders, the gelcoaters, and all others in the process. Anyone who produces poor quality goes upstream to help correct. The amount of WIP basically never changes, and no one gets ahead of schedule. It is basically a self-checking system.

These efforts impacted the facility in a favorable way financially. In 1998, the plant incurred $1.3 million in unfavorable production variances. In 1999, the plant had favorable variances of $140,000 and was profitable. Wow! (War on Waste!)

There was a bright side to all of this. We really learned what not to do in designing boats for manufacturing and assembly. We actually got into the zone of teamwork with the corporate design and engineering team in introducing a thirty-five-foot express cruiser. We all basically listened to each other. They came to the plant and listened to the issues we had experienced on the other boats. They explained concerns they had about the lack of certain quality disciplines in our plant. This was the best model introduction in my career. Design and engineering developed a set of jigs, fixtures, and tooling for the boat. They designed modules from the start and considered the goal of short production cycle time during the development process.

During the development process, we sent several of our key people to design engineering to participate in discussions, discuss DFMA preferences, and understand the boat prior to production start-up. This process started when the boat was just a cardboard mock-up. Engineering built the BOM as they built the prototype and coordinated with the plant BOM Specialist. They scheduled the first two boats worth of materials to show at the plant.

After the tooling arrived, the manufacturing engineering (ME) team followed it and the boat through the production process, assisting boat builders with documentation and familiarizing them with the boat. They created a digital photo album every step of the way to use for reference. They built one prototype—off-line. They then moved the boat into a production line and filled the WIP footprint for it. MEs worked on the line until production learned the boat and took it over on their own.

This introduction was a great success. A solid and disciplined development process combined with teamwork with design engineering will work. Also, good boat builders will make good design engineers great.

10

Components Plant

After six unbelievably rewarding, educating, and long years, I wanted to move on. I found an opportunity to take a sabbatical.

ON SABBATICAL

I wanted to rest and recharge, and spend time with my kids. I also made the decision to pursue a Six Sigma Black Belt certification. I had taken a lot of training over the years and had led many projects, but I still had a lot to learn. So for six months, I got up every morning and took my kids to school, which was fun. Waiting in the drop-off line, and then watching the little rascals carry their book bags into school with all the other little people was a hoot. I then went to the YMCA and worked out, and then headed to the public library to study Six Sigma. I left there to pick up the kids from school and then did "Dad things" with them. It was great. After everyone went to bed, I would study some more. Sometimes, I went outside during the night and walked around looking at the stars. I had forgotten what they looked like. One evening, I got a glass of wine, sat on the patio, and just watched the starry sky for over an hour. To this day, I still randomly head out in the middle of the night to watch the stars.

I also set up camp in our dining room. The dinner table had my laptop and a stack of Six Sigma-related books. Being a "left brainer," I was particularly concerned about the requirement to apply and use statistics. Although I had taken statistics in college, my confidence was low. I bought a cool software program that had hundreds of multiple-choice statistical problems. It would "ding" If you're answer was right, and "dong" if your answer was wrong. Let's just say there were a lot of "dongs." This exercise was fascinating to me because it really showed me how our brains work. When I first started studying, I would work for hours on one problem and get a "dong." I was starting to get concerned. I had already scheduled my exam in Toronto in three months and purchased airline tickets. The worst thing was that my kids were really paying attention. They were fascinated by my studies, and I did not want to fail in front of them. I wanted them to see that if you work and study hard, you can achieve.

Well, the hours and "dongs" began to subside; and after a few weeks, I was working problems in an hour and getting "dings." The problem was that the exam was 200+ questions/problems in 4 hours, or 240 minutes. Oh, oh! I kept at it and before long, I could do problems in just a few minutes. The software was great because if you got the problem wrong, you could click on "Review," and it would fully explain how to approach the problem with a detailed discussion. I had also created a stack of three-by-five-inch cards that was about six inches high, on which I had written questions and answers . I was now correctly solving problems in six minutes that I could not solve in six hours a couple of months earlier.

I flew to Toronto to take the Six Sigma Black Belt exam. I got a hotel room on Yonge Street. There was a Starbucks® nearby. The exam was in three days. I worked problems in the hotel room, and would take breaks and walk up and down the street to get a cup of coffee or eat in one of the restaurants, of which there were many and they were awesome. The best Chinese restaurant I have ever eaten in was on Yonge Street. I think it was called "The Spring Roll" or something like that. I have often thought of going back there just for fun. What a great place. So I studied for three days—morning, noon, and night. I then put down my books and committed to a nice relaxing evening and a good night's sleep.

The next day, I got up, picked up a big cup of Starbucks coffee, and headed for the testing center. I took my luggage with me so I that could fly home shortly after the exam. This was going to be a large sitting for the exam. There were people everywhere, and they were from all over the world. Dark, light, Russian, Indian, Pakistani, Latino. Wow, what a melting pot. It was fascinating. Many of these people looked distinguished and intelligent, and here I was, a boat builder from a small town in Michigan.

After checking our credentials, etc., we entered a very large auditorium-type room filled with tables that were about four by six feet. Each examinee got a table. I picked a table in the middle of the room and placed all my books on it. The test was "open-book." I had carefully tabbed all my books to quickly find certain topics—should I need to. As I arranged my books, a voice behind me asked, "So, is this your first time taking the test?" I turned around and said, "Yes, what about you?" The man said, "No, this is my third time." I asked him what he did. He said that he was a consultant. Great, this was just what I needed to boost my confidence! Not! I have always had more ambition than intelligence, and I was pretty intimidated by this test. I had put a ton of work into preparing for this test and did not want it to go to waste. I also did not want my kids to see me fail the test. I wanted them to see that studying and working hard works.

I took the test. I first answered the questions that appeared to be easy, starred the problems that I knew I could do, and then put an exclamation point by the ones that looked hard. Next, I worked the problems I could. Then I started on the hard ones; and believe me, they were hard! The test started at 10:00 a.m. and finished at 2:00 p.m. They would call out the time every thirty minutes. Talk about time flying. In the blink of an eye, the moderator said, "Fifteen minutes." I finished another problem when the moderator said, "Five minutes left." I guessed at the remaining questions. It was over, and I was tired.

During my sabbatical, I had received several calls from a good friend who worked at very large mega-yacht manufacturer, actually the largest in North America. It was located on the West Coast, and they had built a new facility to build their very large flagship model, which was to be the largest fiberglass production mega-yacht in the world. The company had also obtained a facility to manufacture and supply the interior components to the new facility and two other production facilities. The company needed to transition from having a small cabinet shop to creating a factory that would not only deliver the interior components for the flagship on time and to extremely high quality standards, but also bring back in-house a significant amount of outsourced work. The company had some incredibly talented craftsman and woodworkers, but they needed leadership in setting up the processes required for much higher levels of production. There was also a high level of concern about meeting the interior component demand for the new flagship. In some ways, the company had bet its future on this particular yacht. They were taking no chances on missing the delivery of the first unit.

After discussion with my family, I flew out to the West Coast to talk to them. I was hesitant to even think about moving that far west. My center of gravity was on the East Coast, and it seemed like a far-away place. I had spent a summer in Fort Lewis during college to attend ROTC training. It was a beautiful place—but again, very "far away." I landed at SeaTac and took a shuttle to Boeing Field. The company planes pilot me there to fly me the rest of the way. It was a twin-engine Cessna. We flew across the Puget Sound, the edge of the Olympic Mountains. From my seat in the cockpit, what a view it was. I could see Mount Rainer, Mount Hood, Mount St. Helens, and Mount Baker.

We landed at a small community airport that was literally within walking distance of the main plant. We entered the lobby, and I was welcomed and took a seat. Immediately I noticed how nice and friendly everyone was. They were excited that I was there, and it made me feel great. My friend came into the lobby and greeted me. It was great to see him. He gave me a plant tour, and the plant looked awesome. The boats this plant produced were 130 feet in length and magnificent. Quality was evident and obvious, and it also implied a highly talented labor force. The plant was clean, organized, and they were implementing 5S. I was impressed!

We went back to the front office and into a conference room to meet the founder of the company. Rick and his brother had started this company, and it was growing. It was an awesome story of entrepreneurialism. This plant was also "in the middle of nowhere" and was providing wonderful jobs for the community. It was all good.

My friend introduced me after Rick came into the room. Rick was an interesting sort. He was wearing white coveralls and looked the "engineering type." He served in the role of VP of manufacturing. Rick was a very sincere and down-to-earth person, and earned my respect right away. We talked about the components plant and what they wanted. He explained to me that they had bet the company on the new flagship yacht product, and it was imperative that the delivery of interior components to the first flagship in particular not be a constraint. Failure was not an option. It was obvious that this man knew every square inch of every product and plant. He asked me if I had any questions, to which I offered a couple, and then he asked my friend if he had anything to say. My friend said, "Bill is a good guy, and he will get the job done. I think you should hire Bill and pay him whatever he wants." Rick told me to go look at the plant and let him know if I was interested. I attended a cook-out that night at my friend's beach house and we had beer, and bacon-wrapped scallops and some halibut, both grilled outside under the stars in the cool and balmy Pacific Northwest.

The next morning I was back on the plane flying along the coast of Washington with the cobalt-blue Pacific on one side of the plane and the Olympic Mountains on the other. We flew right by Mount Olympus. As we approached the Puget Sound, the pilot pointed out Port Angeles. It was a small town right on the water. We were as far as west in the contiguous United States as you could go without getting wet.

I arrived at the plant. It was a busy place with quite a bit of work-in-process (WIP). I immediately noticed the components. The benches and areas contained cabinetry, tables, furniture, doors, wainscot panels, and countless other components. What jumped out loudly was the unbelievable level of product quality, as well as the amount of talent the workers must have had. The look of these components was nothing less than stunning. But then, considering that the mega-yachts they went into sold for millions of dollars, some upward of $30 million, it was understandable but no less impressive. They purchased veneer logs, cut into flitches, and glued them to plywood or other substrates. They carefully created "flame" and symmetrically mirrored grain patterns, glued and pressed them, used them to assemble components, and then went through a very labor-intensive finishing process that produced incredible results.

I took the proverbial "ride around the area," snapped lots of photos, took some video to show my family, and headed for the airport. Immediately after I returned back home on the East Coast, I received a call from the head of HR suggesting that I bring my family out to take a look around. So my wife, my kids, and I headed out to Port Angeles. The mountains were beautiful, and the surrounding areas revealed view after view of beautiful sites—almost to the point where you tire of saying "ooh" and "ahh."

The ASQ (American Society for Quality) posts the results of certification exams online. They tell you to start checking after about four weeks. I had put a tremendous amount of work into preparing for the exam and doing my projects. I was uneasy, and I wanted a Six Sigma Black Belt badly. I will bet you that I was logging in five times per day. My family and I stayed in a hotel on the waterfront. After we had settled in for the evening, I checked online for my certification. Still nothing. The three-hour time change kept me up late. I was surfing the Net at about 2:00 a.m. and, once again out of habit, checked the ASQ site. I clicked on my name, and it said "Six Sigma Black Belt" next to it. It didn't even hit me at first. Then I yelled, "Yahoo!" and woke up my entire family. They knew what it was and we all celebrated. Then they suggested that I go to bed. That was a cool moment.

The company made me an offer, and I could not believe how incredibly nice this company was treating me. I had never experienced that type of thing. I had been treated well by other companies, but nothing like this. I accepted a very generous offer and away we went.

LET'S GET LEAN

I traveled to the West Coast ahead of my family to start work. They would follow soon. On my first day, I experienced something that I had never before experienced. When I got up at 5:00 a.m. to get ready to start the day, I got this hollow feeling inside of me. I felt like I was on the other side of the world. I had a bad feeling, which was not like me. I called my wife and told her that I thought I had made a mistake, that this place was way too far away. She told me I was just stressed and to go to work. A couple of hours later, I was prowling around a new Gemba and never looked back.

The importance of top management support is stated again and again in implementing Lean or leading change. Shortly after I started, Rick came to see me. Rick's guidance to me was that, "We hired you to do a job. Do not let anyone or anything get in your way. I am a phone call away if you have any problem." He also made it clear to everyone that he supported my efforts and backed me up. He showed this through both words and actions. Then he turned me loose. This is exactly the top management support that is talked about in all the Lean books. The amount by which a company improves its performance and processes is directly tied to the support that upper management—particularly the president or owner—gives it. Significant improvement in any company is nearly impossible without clear, out front, and unending support of the continous improvement effort by the president or CEO.

The team at the components plant had a lot of highly talented woodworkers. These guys were great. The cabinetry and woodwork they did were incredible—and is to this day. As I said, I had just finished getting my Six Sigma Black Belt and attending Lean 101 training and continued to read and study everything Lean that I could lay my eyes on. I was really getting excited about the possibilities with regard to implementing Lean in this plant.

I began the Lean transformation and set my priorities toward people, process, and quality. Although seldom mentioned, one of the biggest benefits of implementing 5S is the opportunity it provides leaders to build relationships with their people. I started by setting up a communication board with our goals and plan that we would continually refer to. A clear explanation of 5S was part of what was displayed on the board. I explained to the team about the concept of a *visual* factory. One aspect of a visual factory that I feel is important is to being able to see the status of other operations by just looking toward it. So I asked that anything higher than waist-high be removed or taken down if at all possible. There was a large framed area in the middle of the plant with patterns hung all over it—it came down.

As with many work environments, the operators were "dug in." Most personalized their areas. One challenge was that a short time before I got there, every builder in the plant built a tool chest. They were about five-feet high with four drawers, a top, and some sort of bulletin board on them. Here is the thing: nearly every drawer in every one of them was empty! There were calendars, and pictures of cars, deer, and whatever on them. What was worse was that these guys loved them, but they were taking up six to eight square feet of floor space and were not really being used for anything. Further, they were less than three feet from a workbench containing about eight drawers—and again, a lot of them empty. The company had a strict policy about giving away anything to employees. I asked the owner if he would support an exception so I could get these out of the plant without upsetting the team any more than we had to. He agreed to it. So away they went.

The next challenge was getting the guys to sort and get rid of things that were not used anymore, or would probably never be used. They were packrats, but no different than anybody anywhere else. We had many bench-side discussions about whether or not something should go. Some of them were tense, but we managed to get everyone there. We filled a small building with red-tagged items. The toughest challenge was to get the guys to buy in to standardizing the layout of workbenches and drawer contents. These guys were seasoned and talented craftsman, and they saw 5S as someone (that is, MEs) meddling in their areas. But again, we got them there.

This was a tough—and I do mean tough—5S implementation. One operator (Steve) was struggling. He was a great guy and a talented cabinet builder. He was struggling with the rhyme and reason for doing 5S. He took offense to the "before pictures" we took. I take leading people very personally, and I felt bad that Steve was struggling with this. I wanted to help him through the 5S process and get his buy-in. So I did some research and found a book titled *5S for Operators*. I was excited about getting the book. When I got to work, I headed for Steve and said, "Hey Steve, I got a book about 5S for you that I think you will like." He said, "I'll tell you what. I'll read a book if you read one." I said, "Sure, what book do you want me to read?" He said, *How to Win Friends and Influence People*.. Several months later, Steve apologized. I gladly accepted it, and we have maintained a great relationship ever since.

The pride was beginning to grow. The guys gave their families a tour during the company picnic. The same guys who had given me mean glares during the 5S process were bragging to their families and neighbors about doing 5S. I got a real kick out of hearing a couple of them explaining to guests what 5S was. It was a very rewarding day.

So, over a couple of months, we had implemented 5S and opened up the plant in a way that you could see anywhere in the plant from anywhere in the plant. I had begun to build relationships with the team and was getting to know them. Our company airplane parked next to the plant when visitors flew in. I began to notice people passing through the plant often to take in the improvements. That was fun.

We mapped out the process on a whiteboard in the office. It was the center of many discussions with the team. Debates ensued on how product should run through the plant. I explained what flow was and how important it was, and asked them to consider that. These activities yielded a future-state layout that everyone was excited about. Literally every bit of it came from the team. However, it involved relocating equipment, the services of electricians, re-routing dust collection ducts, and setting up equipment. I wanted to do a Kaizen event to get it done, but it was a lot and I was concerned about how doing all this would impact production. The team said not to worry, that they had a plan. They scheduled the following week's work to be done by Thursday at noon. At noon on Thursday, people came out of the woodwork from all over the company. Electricians came up from the flagship plant and brought wire, Maintenance began re-routing ductwork, and the leadership team and some employees began to rearrange the shop. By Friday afternoon we had a new plant layout, which made flowing product through the process much easier.

So as we continued on, and as a result of our scheduling process, we were getting out of balance between work centers in the plant. One area would be stacking parts and assemblies while the work center needing them was swimming in them, and the next work center after that was starving for work and the operators were standing around or passing time. It was not working. We decided to develop standardized materials kit carts across the plant. We also established an authorized number of completed kit carts for each area. Once they filled them, they had to stop work and go help another work center catch up and remove some of their carts. Once they had a couple squares open up, they could go back to work until they filled them again. On the side of each cart, we put a laminated card that was red on one side and green on the other. Each area's team leader would look at the routing sheet on the carts and put the red side of the card out if it was late, and the green side if it was on schedule. Work centers always worked on carts that had red cards on them before the green cards.

Applying veneer is one of the most important and key processes in interior components producing for mega-yachts. Logs of rare wood were purchased for thousands of dollars and sent out to be cut into veneer, often overseas to other countries. The logs come back in a package of "flitches." The process of applying veneers to create the premium or superb wood finishes expected by someone paying in excess of $30 million for their mega-yacht is challenging and requires a lot of talent.

Samples from the logs are laid up and presented to customers. Once they find one they like, the veneer team sorts through the flitches in the log and selects veneer pieces that complement each other and will create a consistent look in each area of the boat. The color, grain, and finish must be consistent within each room and—to a point—throughout the boat. Another challenge to this process is that many of the rare woods change colors when exposed to light. Great care must be taken to make sure that all the components' exposure to light is controlled and consistent. Leaving a beautiful piece of veneer out in the light will change its color, often to a point that its color is no longer in harmony with the rest of the components in that room. This can yield some costly and disruptive rework.

One of the most desirable features in interior wood on mega-yachts is called "flames." The grain pattern needed to create flames requires wood from the log where there was a trunk or forking type branch located. These cuts of veneer are used to create the flame patterns. This pattern usually produces a highly figured V-grain that is beautiful and highly desired.

Another popular veneer feature is called "book matching." This effect is achieved when two sheets of veneer that have been cut from a log in consecutive order are joined to create the appearance of mirror images of each other. Once the size of the wood component is known, someone must sort through veneers to find two pieces large enough to cover the component. These pieces must be defect free and of the highest quality.

Next, the veneers are glued to a plywood type substrate and put in a heated press to achieve proper adhesion. Veneers must have a very straight edge on them to be effectively joined together. A machine called a "cutter" is used to achieve this edge. There are different techniques for joining the veneers. Glues, pins, and tape are used to fasten the two edges together and create the effect. These operations can be done by hand or by machine. The proper glue is critical because the veneer depends completely on the glue. Depending on the application, one or both sides of a substrate may be veneered. Generally, when veneering, if you veneer one side, at a minimum you must glue something else to the other side. Otherwise, warping issues, etc., can occur. This is the same concept as that of alternating nuts while changing an automobile tire. After gluing, the veneers are sanded, usually running through a sanding machine, a process that must be very accurate. Once the veneering process is complete, the veneered panels are used to fabricate interior components. This may involve manual cutting operations or CNC machine cutting operations.

So, the operations involved in veneering were sorting, cutting, layout/grain matching, taping/assembly, seaming, gluing, pressing, trimming, finish sanding, routing, and calibration of materials to proper thickness. We worked with the team leader and operators in the area and determined the operations, determined the optimal worktable size, and the optimal number of worktables; we worked to minimize the amount of travel during the process, and we set up the equipment in a U-shaped cell. However, we did consider organizing the work in a line, but ultimately decided against it.

One day we received a memo from the founder of the company. He explained what he called the "standard target." It was his take—and I think an interesting one—on delivering value to the customer, noting that value is the only thing a customer will pay for. He sent a diagram that had cross-hairs on it. Our job, he said, was to hit the standard right on. If we give less value than the customer is paying for, he will be disappointed. If we over-process and give more than the standard, which is what the customer wants, he is not willing to pay for it. We had some employees who were working to make every boat better than the previous one. The idea was that if the previous boat was exactly to standard, which is exactly what the customer was willing to pay for, then going beyond that was waste.

During the time frame of process improvement and applying Lean, we also focused on our internal customers. This is always an important area to understand. We visited the plants that installed the components we manufactured to understand how we were doing relative to their expectations of on-time delivery and quality. We started going to all the plants once per month and had weekly conference calls. It was tough at first, but we were able to improve.

It was during this assignment that I met a man that I would call a genuine Sensei. He was a small man who earned big respect from all of the employees. He started a boat company many years before where he worked with many of the employees. He also taught boat building at a local school. He was an expert in boat building and woodworking. Whenever the guys would have a question about boat building, they would go to him. I often saw one employee start a conversation about something with him and before you knew it, it would turn into a class of six to eight people and every eye and ear was on him. If he said it should be done, it was done.

The experience of setting up the components plant was one of the most meaningful experiences that I have had. Every group or situation is different. I felt that I had hit the sweet spot with change and improvement leadership on this project.

The team did an awesome job at 5S. The plant was clean and organized, there was a place for everything, and everything was in its place. We opened up the plant so we could visually manage the operation. We could see work-in-process (WIP) and all work centers from anywhere in the plant. The materials carts managed with Kanban principles worked. You could see clearly which areas were backed up from anywhere. They provided a simple, yet powerful tool for maintaining the balance of WIP. The employees in the plant also knew when to shift resources and help another area—it did not even require a conversation. Each area had a "whiteboard" where the schedule and any other key information were displayed. The other side was a dry-erase board on which they would note any issues that were impacting their ability to achieve schedule or quality. As leadership passed by an area, they observed the issues on the board and took action to help. You could go into any work center in the plant and within twenty seconds determine its status with regard to quality, schedule, or challenges without the aid of a computer. Where have we heard that before?

We had executed some exciting and powerful Kaizen events. The 5S implementation was one of the best deployments of this philosophy for managing a workplace that I have ever experienced.

The Kaizen event to change the layout of the plant to improve flow was one of the most exciting events I had ever participated in. To see other employees coming in from other areas in the plant to help—electricians, forklifts, people to help make and install new duct collection ducts. Seeing the excitement in the employees participating was an incredible experience. This was flat-out fun! Setting up work cells for veneer processing and edge banding created real benefits and an added sense of pride in giving the boat builders a better work environment.

In less than one year, the components plant experienced improvements in product cycle times, flow of product throughout the operation, and quality, and a state safety inspector said the plant had the best equipment guard use and safety discipline that he had seen in his fourteen years in that role. At the end of the day, all components for the first flagship tri-deck motor yacht were delivered to support successful completion of the first unit as requested.

As I close out this transformation account, I am reminded of a good day I had one Saturday during my first few weeks at the plant. I was working lots of hours during the week and looking for a house on the weekend in preparation for bringing out my family. One week, my friend Matt mentioned to me that Rick, the founder, was going to be in the plant on Saturday and he was looking forward to seeing me. So, Saturday morning I headed down to the flagship plant and found Rick. He was on a forklift with another guy, assisting him in pulling the deck for the flagship from the mold. It was just Rick and this guy.

I have always been proud of starting out in the factory as a boat builder, so I quickly rolled up my sleeves and started helping them pull the part. I grabbed some chain and suggested the location to hook it. I had not ever pulled a part that big from a mold, but I had sure pulled a lot of smaller ones. The three of us worked together for most of the afternoon getting the deck out. Toward the end of the day, Rick and I were outside the back of the brand-new waterfront plant. As I looked at the beautiful orange sunset over the Puget Sound, I commented to Rick, "It just doesn't get any better than this—beautiful plant, beautiful boat, on the beautiful Puget Sound." We shook hands and I went home. That was one of those moments that I will always remember. Manufacturing has been good to me.

11

Lean Transformation #2

My next opportunity came in the form of an offer from a former mentor to be the vice president of quality and process improvement for Genmar's Sarasota operation where they produced the Wellcraft®, Hydra-Sport®, Scarab®, and ultimately the Sea-Swirl® brands. I accepted the offer. We had family ties in the Sarasota area, and I really like Florida. I was particularly excited about this opportunity because my primary responsibility was to improve processes by leading a Lean Six Sigma transformation. Continuous improvement and leading people have always been my passion. Finally, I was going to focus solely on my passion: improving processes.

PROCESS IMPROVEMENT AT WELLCRAFT

I arrived at the Wellcraft plant and was ready to go. Anyone who knows me knows that I have, let's say, "high energy." I am passionate about manufacturing and boat building, and I am always, always in a hurry to make things better. It was a new Gemba, and I was fired up. Twenty-five-plus models, work-in-process (WIP) everywhere, quality defects everywhere, waste everywhere I looked. Eeehaa! A process improver's dream!

The Wellcraft brand had a long and proud history. The company had provided jobs for hundreds of people for many years. It had a history of great product such as the Scarab and Portofino®. Many famous celebrities had walked the floors of the site over the years, including Hulk Hogan, Emerson Fittipaldi, Don Johnson, and even Ringo Starr. The company had a hugely talented labor force and many industry veterans in management. I was proud to join this team. However, for all of its history, the company was now facing significant challenges.

One thing I appreciated about Genmar was the way they treated their people. They respected and appreciated the experience people had in the marine industry. I had watched many incredibly experienced managers who came up through the ranks in boat manufacturing leave US Marine®, and it bothered me. These were exactly the kind of people who are valued so highly in Lean Six Sigma. For a period of time at Bayliner®, having boat-building knowledge was almost a black mark on you. I am sure that some would disagree with me, but all I will say is: Take a look at where US Marine is today. It seems to me that the more people with industry knowledge who left the company, the worse the company performed. The more outsiders with pretty resumes who came in, the worse and less profitable they were. On the other hand, Sea Ray® seemed to appreciate and hold onto more of its talent, and I believe that is a key factor in how well they have fared the latest economic downturn.

The Sarasota operation was not performing well in the areas of quality, cost, and schedule at all. The processes were not being controlled, and the plant lacked strong leadership and an understanding of Lean Six Sigma, effective process improvement, or problem solving using root cause analysis. There was WIP everywhere, WIP with nothing being done to any of it. There was also an excess of management in the plant; managers, directors, and supervisors were everywhere. I think the upper management had struggled with getting to the real issues and just kept throwing people at it.

This was one of the most political environments I had ever seen. There were cliques, alliances, and rivalries, and these guys had all the office-politics skills. *Big Brother* and *Survivor* had nothing on this operation. That is usually a problem for me, as I have about as much chance of being a good politician as I do of becoming a professional golfer. Their skills included not giving away their position in a meeting, never disagreeing with anything, agreeing with everything but yet go back and not do it, collusion, keeping a straight face no matter what you are thinking, and the ability to provoke. The weird thing is that I have read countless books on Lean and Six Sigma and have never come across a chapter on these skills. I understand the need for some political skills, but manufacturing has no place for politicians. You cannot have politics *and* teamwork. You get one or the other, and it takes teamwork to turn around a manufacturing operation and sustain high performance. I went to work.

Organizational politics are a hot button for me. I cannot tolerate politics or politicians. I am OK with professionals who have political skills, but not people who exist primarily on politics. I will also admit that I have gotten my "clock cleaned politically" a couple of times. I recently brought up my disdain for politics on my blog. I was surprised by some of the responses from relatively accomplished professionals. "Leadership is politics," one person wrote. Really? Is putting one's personal agenda ahead of that of the team or organization leadership? Is working against a teammate or not offering assistance to them because you are working against them politically leadership? Is withholding information that would be helpful to a teammate leadership? One last question for those of you who think politics are OK: If politics is leadership, and just fine, then why are there never any seminars that teach political skills? All politics come at the expense of teamwork, and teamwork is essential in a Lean organization.

Let me say that I view politics as a symptom. It is not a problem; it is a symptom of a lack of leadership, clear direction, teamwork, and accountability. Politics grow like mold when a refrigerator is left open. So, as I describe and talk about this situation, it is not about the individual character of these people; rather, it is about the nature of people. It is about the natural evolution of human behavior in situations without effective leadership.

Just recently I learned of an organization (that I am highly familiar with) losing a plant manager and his complete staff at a key plant. I do not know if they chose to leave, were asked to leave, or some of both. During my time at that organization, I had noticed that the plant manager had become increasingly political, and ultimately excessively political. It was pointed out to management but it was not addressed. The speed of the team is the speed of the boss. This manager's staff followed suit and, in turn, became political. Negativity and a lack of teamwork became the norm as the team put its energy into political activities as opposed to continual process improvement and teamwork. This level of political activity was a direct result of a lack of accountability and effective leadership from above. Evidently, the team's behavior morphed into something unacceptable because all of a sudden they exited the company. In this case, politics and a lack of leadership and accountability ruined a team of highly talented people. Politics are costly—and also highly contagious.

In my experience, an important responsibility of all leaders is to quell politics, and they must lead by example. The leader should advise that politics will not be tolerated. Politics erode teamwork, and I do not know of a way to lead an organization forward without teamwork. OK, I am finished.

I digress.... Coincidentally, a few years earlier, the president of Wellcraft had asked me to spend a couple of days touring the manufacturing operations and give him an assessment of what I saw. The first thing I noticed was a huge amount of WIP. There was fiberglass everywhere—and I mean everywhere. Now that might have been all right if it had operators on all of it and it was on a balanced line operating to Takt time. However, only about half at best of the WIP was being worked on. I asked the vice president of manufacturing how things were going. He said, "Awesome! We are hitting really high efficiencies. We actually broke 100 percent efficiency a couple of weeks ago!" The problem was that, according to their reports, they were. It was evident that the labor standards being used to calculate labor efficiency must have been inflated.

I then began to check out the quality process. There was a lack of quality controls or a process in the first area of the process, that of fiberglass lamination. This was ironic because fiberglass defects were one of the largest categories of warranty expense. Equipment did not appear to be maintained, and there was no sign of calibration. In the final process (i.e., assembly), there were quite a number of QA auditors verifying product to quality standards at the end of the line. Lists of defects were being created, with constant requests to upstream departments to come out and address them. What I can say was that the people in the process were working very hard, and the operators were some of the most talented boat builders in the industry. There clearly was no lack of talent. There was a clear absence of disciplined processes and leadership.

The general finding was that the WIP most certainly indicated that labor costs were out of control and that the majority of the quality team was focused on areas other than the area generating the highest warranty costs and customer dissatisfaction. I gave him my assessment and moved on.

OK, back to my new role. As VP of quality and process improvement, I had direct reports of the director of customer service, director of manufacturing engineering, quality manager, and their associated teams. After a couple of days on the job, I sent out a request to get together as a team and start the process of moving forward. You know, the people, process, quality thing. So, I sent an e-mail to the team, asking them to meet in my office the next morning at 9:00 a.m. Well, one guy said he could not make it. He had a prior commitment. OK, somebody help me out here. Can you imagine in your life that your new boss asks you to meet with him and you tell him, "Sorry, I have another commitment"? Oh, boy. So, being the reasonable guy that I am, I sent a new e-mail that said, "OK, how about 1:00 p.m.?" Another guy says, "Sorry, can't make it." OK, I sent yet another e-mail out suggesting 6:00 a.m., and sure enough, somebody could not make it. They had another meeting. OK, so I sent out a meeting request and asked if anyone had a commitment at 4:00 a.m. Awesome, all their calendars were wide open! I scheduled the meeting for 4:00 a.m., and Joe, our manufacturing engineering manager, said he would bring the donuts. Ah, leading change.... And I never again had a problem scheduling a meeting.

The rumors started. Politics were everywhere. People would tell me that others were saying they were going to outlast me. Outlast what? Improving quality and processes? I would call meetings or set up brainstorming sessions to solve a problem, and we would hammer out a solution using their ideas. Everyone would agree to take corrective actions, and they would leave and not do anything differently. When I followed up with them and asked them how they were doing with the plan, they would give me a blank stare. This was not for the weary. People knew that when things are not going well, then sooner or later someone will most likely exit the building. That brings out their survival instincts. That is a big part of what politics are, a strategy for outlasting someone or something else, and resisting change. Regardless, as long as politics at this level are occurring, there will be no effective improvement activities, implementing Lean, or team building. The appropriate thing to do is to drive forward and try to work with people. You drive forward by beginning to measure the process, establishing process standards and standard work, and then setting goals for the team to drive toward. Then teach the team process improvement techniques and problem solving, and help them meet or exceed their goals. Follow that by increasing accountability for adhering to standards and making progress toward metric goals. This is tough stuff but the most challenging and rewarding type of leadership situation.

I am a "War Room" type of person. I like to take a room and line it with whiteboards and bulletin boards, even on the back of the door. Brainstorming with people together to solve problems, debating solutions, or applying analysis is unbeatable. It is so effective, and it is enjoyable. My office is a tool, not a status symbol. I like it in order too. The company had an office known in the industry as a "Taj Mahal." It was built in the 1980s and had palm trees and fountains in the lobby, a sauna, and private bathrooms. It even had a workout room long before they became popular. It had a patio out back with a miniature lake that they used to fill with boats for dealer events, etc. The top floor was the ultimate! Sales, marketing, and engineering were all located there. They were pretty high-end offices.

As the vice president of quality and process improvement, I was given on office on the bottom floor. Hmm. Well it was dirty, very dirty. I came in over the weekend and personally scrubbed the walls, ceiling, floor, and all the furniture. I put up dry-erase boards, corkboards, and motivational stuff. It was a place to get things done. I even went outside and washed the windows. I cannot stand a messy office.

The next week we hired a man by the name of Bill Carreira as vice president of manufacturing. He had written a book titled *Lean Manufacturing That Works*. To this day, I consider it the best Lean book I have read, and I have read a lot of them. In hiring Bill, the company hired someone who turned out to be a personal Sensei and mentor to this day. Bill immediately taught me something and saved me from making a mistake. I invited him into my office. He said, "Come on, I want you to come out into the factory with me. We're going to have some fun; we have work to do and we have to be in the middle of it." I said, "Bill, I just spent the whole weekend setting up this office. I need a cave to get away from it all." He said, "No you don't. Come on, you and I are going to have some fun." He and I went out to the plant and into a little office up near the ceiling where we could see the factory. I was not out there an hour and I realized this was going to be awesome. Bill immediately set up a coffeepot, and we went to work. I had never worked in an office like that before. We hired a director of materials, and he came right in there with us. The communication was so awesome. You could not help but know what was going on, even if just by osmosis. We had two vice presidents and a director in a twelve-by-twelve-foot office, and we worked from folding tables. It was the coolest office environment in my career to this day—and I do not mean air-conditioned office, as this office was just under the ceiling of a factory in Florida.

So the remainder of the plant reported to Bill, including all the production managers and supervisors, etc. I immediately gained an enormous amount of respect for Bill. He is really difficult to describe. He is extremely intelligent and perceptive. He understands Lean, and he is incredible at value stream mapping. He calls it a "baseline." He and I hit it off immediately. They called us Bill and Bill. I am high energy and Bill is laid back and intellectual. I would give motivational speeches bouncing all around, and then Bill would sit back and ask, "So how's it goin'?" Bill also had an incredible sense of humor. He uses sarcastic humor, which I like. Humor is a great stress reliever and lubricant in life.

We became a team, a solid team. This drove the middle managers crazy. Their goal was to come between us. They tried every conceivable angle to drive a wedge between us. But Bill and I just kept working and sharing our passion for fixing things in manufacturing. Anyone who wanted to talk to one of us, talked to both of us, because we sat right next to each other.

They complained about that. We did not discuss anything personal with anyone in front of each other. But if it had to do with manufacturing or quality, you talked to both of us, or maybe the three of us when the director of materials, Glen, joined us.

Some of the people had long ties to other people at Genmar, and they tried to take advantage of those ties to get around Bill and me. But we were solid in our focus to turn things around as a team. We had to lead by example, and we stood shoulder to shoulder. If Bill and I did not work as a team, how could we expect everyone else in the plant to do so?

Lamination Process Improvement

Now, being the passionate scholar of Lean manufacturing that I am, I was totally pumped to be able to team up with Bill in doing a baseline or mapping exercise for the lamination process. We established a team of laminators, boat builders, supervisors, and others from materials, quality, and manufacturing engineering. The majority of the team were hands-on operators from the process.

We followed a process that went something like this:

- Day 1:
  - Walk about
  - Discussion
  - Current state map
- Day 2:
  - Finish current state map
  - Spaghetti diagram
  - Space utilization
  - Work sampling
- Day 3:
  - Current state complete
  - Team overview
  - UDEs (undesirable process elements)
  - Root cause/brainstorm
  - Develop future state

At the end of each day, we would bring in upper management and let the boat builders brief them on what they accomplished that day, and what they had learned. It was good. Bill is a "master" at this. I learned so much from him. The people really enjoyed this approach. We ended with a future state and proceeded to build on it over the next three years.

We continued to work together. Then the rumors started that both Bill and I would be gone in six months. One boat builder in particular was particularly vocal about his feelings. He proudly proclaimed that he would be here long after we were gone, and he was not about to do anything we asked. Now mind you, we were smiling everyday and getting to know the people, and applying some very positive leadership. On the other hand, we were slowly starting toward changing the culture, and this culture was deeply engrained. Let me be clear, however; there were a lot of great things in the culture too. They had some great boat builders and very good employees. However, there was a sizeable group of people who were resisting change, our leadership, and in fact they were working against the effort.

The politics continued. Daily efforts to get between Bill and me. Managers continued agreeing to get something done and then not doing it, and there were more comments about outlasting us. There was one manager who, whenever we would call a meeting to solve a problem, he would ask, "OK, who's gonna get that done?" He would do that every time something needed to be done, and usually someone would say, "I will"; but the problem was that the manager should have been the one to do it. He was slick. Heck, I even volunteered to do things he should have been doing! So one day we were in a problem-solving meeting and something needed to happen. He right away asked, "OK, so who's gonna get that done?" I could not take it anymore, and I said, "You are." And so every time he ever asked that question again, he got the same response from me.

We were not making the progress we needed to. It was a bunch of games. I was starting to grow impatient with the situation and had made some comments to Bill about it. I was suggesting that we needed to do something to gain some forward momentum. Actually, I was not suggesting. I was wearing Bill out. I would ask him daily if he was ready to get going. I did it in a friendly harassing kind of way, but I was pushing him. At the same time, he needed to be comfortable in moving forward, as we were a team. It was about 5:30 p.m. on a Friday evening at the end of a challenging week. Bill and I were in the hot seat, and we participated in a conference call with Genmar on a daily basis, and they were pushing us for results. I came into the office from the plant floor and sat down. Now you have to know Bill, but he was standing there with a big mug of coffee. He looked at me with this really serious look and said, "I think you and I need to closely align ourselves and start moving forward." I smiled and said,

"I thought we were!" And then I noticed how serious he was. Being me, I said something like, "So you are saying that you think we should start to do what we were hired to do?" He said, "Yes, I am, and we have to get some teamwork going, and some people need to leave the building." I said, "I couldn't agree more." Now Bill is a chain smoker, so he says as usual, "C'mon, go for a walk with me." So we went to the smoking area and while he was smoking, we made the decision about some people. I asked him if he was ready to do it immediately. He said, "Yes." So I called the president and asked him if he had a few minutes. He asked me if it was important, and I told him it was. He told me that he was on his way home from work, but that he would turn around and come back to the plant. We could meet in the front conference room.

We were waiting in the conference room when the president came in. He asked us what we had on our minds. Bill explained that we were not moving forward, and that the people dynamics were not what they should be. People were not taking us seriously, and we needed to do something. The president looked at me and asked if I agreed. I said, "Absolutely." So he asked me what we wanted to do. I said, "We need to let the assembly department manager go." He said, "Frank?" I said, "Yes." Bill explained that Frank was not supporting the 5S process, working to sustain progress on the lines yielded from our Lean analysis, or any of the other improvement efforts. He was simply going through the motions, it was not working, and he was spreading negativity. The president said, "OK." He then asked about our plans for the weekend. I said, "We're not done." He gave us a look and then asked, "What else?" We advised him that we intended to remove some others. He asked, "Who?" We advised him: the lamination manager, two lamination supervisors, the quality manager, and a sub-assembly department manager. We had also just let the director of manufacturing engineering go the week before. The president was pretty taken aback. However, he supported the decision. I give him credit to this day. I know that was tough to do. On a personal level, he liked those people and he would not be a good leader if he took this lightly.

There was nothing personal in any of this. I will take this opportunity to say that none of these people were bad people. None of them were. They all had a great deal of knowledge and experience. Again, all of this was about the nature of people, not the people themselves. They were people who got caught up in a situation. To my knowledge, they all found other good opportunities with other companies, and my guess is that all at some point were happy they moved on.

The amount of work we had to do was immense, so we decided to bring in some Lean practitioners from outside. They did time studies and taught others how to do time studies. They set up Lean metrics on the lines for quality, cost, schedule, and downtime, and they set up daily huddles to communicate and resolve issues. They engaged manufacturing engineering to help fix product issues, or develop jigs or fixtures. They sustained the 5S process, set up standard work, set the WIP footprint, and balanced the lines according to the Takt time.

One day in the office, Bill was telling me that his publisher (AMACOM) was interested in him writing another book, and they wanted it to be about Lean Six Sigma. I told him that was great. Then he says, "Hey, you have a Black Belt and I don't. Why don't you team up with me to write a book? We'll have some fun." So we did. We co-wrote *Lean Six Sigma that Works.* It was a tremendous experience. I challenge the use of the adjective "fun" when it comes to writing a book. I would go with "rewarding"—highly rewarding. Thanks for a great opportunity, Bill.

During all of this, I noticed Bill's cell phone ringing more and more. When he first started, he would tell me about calls that he got here and there and how it enticed him. He got a call from a huge mining company in Peru. Someone there had read his book and wanted him to talk to them about a consulting arrangement. I think one day he got a call from PepsiCo®. I could tell that he missed Lean consulting. And although we were applying Lean tools, we did not have solid top-level support. We had support, but not at the level he wanted and felt was necessary. Then one day he announced that he was leaving. I was really disappointed, as I really enjoyed working with Bill. He wrapped up, and we have been friends ever since. He has been a significant and influential person in my professional life, and one of the most interesting people I have ever known.

After Bill left, I moved into the role of vice president of operations and quality. The transformation and turnaround remained my top priority. I hired a Lean Six Sigma manager to assist with our efforts. His name was Carl Arens. Carl learned his Lean skills at Pratt & Whitney and Johnson Controls. These companies were both ahead of their time in deploying Lean. Carl was a relentless project manager. He pushed project teams hard while teaching them Lean tools and inspiring them. I learned a lot from Carl, and a great deal of the success I am describing is a result of his efforts. We both shared limitless passion for all things Lean. We debated Lean topics as if our lives were on the line. People would look at us and shake their heads. We made a good team and remain close friends and fishing partners to this day.

We focused our transformation strategy on the following:

- We applied a process focus to the strategy; there were seven primary processes:
  - Sales and operations planning
  - Materials and purchasing
  - Small part lamination
  - Large part lamination
  - Sub-assembly operations
  - Upholstery operations
  - Final assembly operations
- Our course of action went something like this:
  - We chartered projects focused on sales and operations planning, the "front-end" processes.
  - Next, we implemented 5S in all the facilities.
  - We then did value-stream maps, Lean time studies, created standard work, created balance lines or areas, created metrics and displayed on dashboards, and implemented daily accountability meetings.
  - We chartered and executed several waste reduction projects in key areas.
  - We implemented a companywide quality process. We practiced "train and do" and combined training with the execution of every project. When we mapped, we had training on mapping. When we needed to identify waste, we trained on the seven forms of waste and then used the training to identify opportunities to eliminate waste. We taught people how to use Pareto analysis, create Affinity diagrams, how to conduct Lean time studies, and assist in line balancing. When we implemented 5S, we trained people at every stage.
  - I created a Lean training program that I called "10 minutes of Lean." Every morning at the leadership production status meeting, I taught a topic from the Lean body of knowledge. For each topic, I used a specific example from the current production environment to aid the participants' understanding. I gave quizzes. One part of the training that really stood out was that the president of the company sat in the front row, participated, and took the quizzes. This was a great example of the type of executive leadership that is so critical to effective Lean transformations.

- We implemented 5S with vengeance everywhere. Area by area, and line by line, we brought in the teams and trained them during each step and then went back to the shop floor and did it. The teams created shadow boards, labeled spaces, and removed unused inventory and debris from the areas. We developed a weekly 5S/safety audit by including our critical-to safety checks. Next, we eliminated the safety committee, which had just turned into a free lunch and complaint session. To take a different direction, we created a team of 5S auditors who volunteered. Every week, every area received a 5S/safety audit by auditors from other areas. The audit was scored, and everyone was challenged to achieve a 70 or higher score, and then higher goals after that. The leadership team and myself met weekly with teams that did not achieve the minimum score to brainstorm with them to determine what could be done to raise their scores and offer assistance to them. Each month, the highest-scoring area received a pizza lunch, and the teams with the highest annual score were invited to a catered banquet. This was hands-down the best 5S/safety effort I had ever been part of. The support and energy of the employees and leaders was amazing.
- To manage communication and ensure performance and focus, we assembled four-by-eight-feet horizontal "whiteboards" and placed one in every area in the plant. (See Figure 11.1.) Each

FIGURE 11.1
These whiteboards were placed throughout the plant.

> whiteboard contained simple metrics related to quality, cost, delivery, and safety. A corrective action log for each focus area was included, noting assigned tasks and the name of the person it was assigned to. Under the quality section, the first time pass (FTP) scores for that area were posted with the top defects indicated. In addition, direct customer feedback regarding product issues, or work performed in that area, was posted. The 5S/safety audit scores were displayed, again with a list of corrective actions. The delivery area showed a copy of the current schedule and the on-time performance month-to-date for the area. Finally, the labor efficiency and materials usage variance for the area was indicated. On the end of the board was a space allocated for issues such as quality defects from other departments, lack of or issues with product documentation, or any issue causing downtime.

The team leaders were required to meet with their teams at the start of each shift and review the whiteboard focus items, address any issues, and review the workplan for the day. I along with the leadership team met in front of every whiteboard with the team leaders from each area at least three times per week. The team for the area would review the status of quality, cost, delivery, safety, and issues for their area. Leadership representatives from quality, purchasing, design engineering, manufacturing engineering, and each key process department were present. The area team leader directly engaged the other areas about issues they needed support in resolving, and a name was assigned to the issue along with a commitment date.

These efforts took a lot of hard work and commitment, but the focus generated from them was incredible. People who missed commitment dates were challenged professionally and relentlessly. If necessary, I would go to the executive team for resolution, but we never let up. The corrective action for the issues indicated that the whiteboards created a challenging workload to get them. However, each and every issue was directly impacting the operators' ability to do their jobs effectively. Although I would not say that any of the meetings were negative, they were not always fun. Tough issues were identified, assigned, and challenged. As in any manufacturing environment, work was never in short supply and going to a whiteboard meeting might mean that you return to your desk with three more things to do. Overall, the team was outstanding, however. After all, this is the world of manufacturing!

I am going to take this opportunity to review some of the projects we chartered and executed during this transformation. For most projects, we used the DMAIC (define, measure, analyze, improve, control) process and most of what I call the Lean tools under the phases of DMAIC. We used value stream and process mapping extensively, along with metric development and PDCA (plan, do, check, act). In some cases we used FMEA (failure modes and effects analysis), and on one or two occasions we did DOE (design of experiments) and ANOVA (analysis of variance). 5S was key and executed with a high level of discipline. The concept of first time pass (FTP) was a key focus. We continuously measured and tracked the percentages of tasks that were accomplished correctly the first time.

Mold Care Project

Fiberglass boats are manufactured in fiberglass molds. To manage the demands placed on the materials in a mold during the lamination process, molds are made from a higher quality tooling gelcoat, and unique fiberglass cloths, etc. Great caution must be used in handling and protecting molds during the production process.

The surfaces on all parts coming from a mold are mirror images of the mold in which they were laminated. If there is a scratch on the mold, there is a scratch on the part. If there is a crack in the mold, a defect in the shape of the crack is left on the part. Obviously, if there is a rough spot on a mold that leaves a rough spot on a part that requires five hours of rework, then if twelve parts are pulled from the mold, sixty hours of rework have just been generated…thus, the importance of maintaining molds—or "mold care."

All fiberglass boat manufacturers have mold maintenance departments, and most follow a disciplined process for maintaining them to minimize rework. A plant can fix a defect once on a mold, or once on every part. The challenge is that tooling is expensive, and it must be pulled from production to be maintained. If the lamination cycle time of a boat is eight hours and it has sixteen hours of mold repair, it will reduce output by two boats for the week or month, etc.

This plant had significant issues with its molds. Compared to the other companies in Genmar, its mold maintenance costs were the highest per standard labor hour produced in the company.

We chartered a project to improve this process and achieve better results. (See Figure 11.2.) There are many things in the overall process that can generate increased mold defects: the quality of the tooling itself, handling of

FIGURE 11.2
A photograph of talented people working together to improve a process.

the mold during transportation, pressures and impact put on the mold to get the finished part out of it, the wax or releasing agent applied to the part, and some of the chemicals such as gelcoat applied to the mold. This was a tough project. Genmar was a holding company, and each of the companies had different part numbering systems, different tooling standards, different tooling production processes, and different accounting processes. This all made using data to compare companies or infer root causes very difficult.

The plant's total cost of mold care in the business unit was the highest of all the units in the company, and was 37 percent higher than the company average. The scope was to determine the drivers and develop initiatives to drive down overall costs. There was a note to compare the measurement systems across all the business units for consistency. The top five cost categories included salaries, fringe benefits, overtime, tools and equipment, shop supplies, and mold repair rework. This project did not achieve the results we had hoped. The company had neglected proper maintenance of its tooling for years, and the quality of the tooling produced internally and purchased externally was poor. The project team concluded that the costs of maintaining tooling would most likely only be brought in line as it was replaced with new, higher-quality tooling going forward.

Cultural issues within the organization, and with tooling suppliers, also impacted the management of tooling. Again, the company manufactured some of its tooling internally and purchased externally. The product development and engineering (PD&E) department was responsible for producing tooling internally. The PD&E area did not have a formal quality process for producing a tooling, which is a very expensive asset. Accountability and verification were minimal. Tooling gelcoat shelf life was not always managed, or was questionable at best. The gelcoat spraying equipment was not consistently maintained, and evidence of proper calibration was absent. The use of mil-gages to control material thicknesses and other process controls was inconsistent. The lack of quality controls and standard work for tooling production yielded poor-quality tooling.

There was a lack of understanding of the importance of process control in the tooling production area. The basic critical-to-quality specifications for tooling include gelcoat thickness, absence of porosity or contamination, gloss, and the absence of geometric angles that will keep parts from scuffing or sticking when a part is pulled from it. Each of these critical-to-quality specifications could be measured. Although ultrasound and gloss meters were available, they were not being used.

A process where production would "accept" tooling from either PD&E or a supplier consisted of production and quality representatives evaluating the tooling and marking up any defects for which PD&E or the supplier would rework and then production or quality would review and/or reject. This process would sometimes go through several cycles. It was ridiculous. Instead of the people responsible for creating a quality product measuring and ensuring critical-to-quality specifications were met, a "cat-and-mouse" game was played. To add insult to injury, tooling is almost always produced for a "new model," which always promised new and exciting features and fresh product, so there is a lot of pressure to get the product into production as opposed to getting the product right. This was the proverbial case of functional silos between areas.

Once this tooling makes it through the cat-and-mouse game and into production, another challenge occurs. Tooling maintenance calls for tooling to be "touched up" to address minor issues every so often, and then majorly reworked at certain points. This maintenance requires pulling the tooling from production, which can be a challenge when a sales backlog exists. Sales resisted pulling the tooling when it needed to be maintained. If it is not pulled when a major rework and repair cycle is due, the tooling begins to degrade at a higher rate. The longer it goes without proper repair, the worse it gets and the longer it will have to be removed from production. During this time, the cosmetic rework from tooling defects on product produced in the poorly maintained tool begins to soar, and so does the cost associated doing the rework.

The project team worked incredibly hard on the project and did reduce the amount of defects, yet the results fell short of what the team hoped to accomplish. The results of the project did optimize the tooling management process relative to the constraints that were not eliminated. Still, at the end of the project, we could not clearly determine why the cost of mold care per standard labor hour so much exceeded that of the other facilities. The variability between the environments and markets for all plants in the comparison created confusion. Companies were operating at different capacities relative to their break-even points, along with the differences between accounting practices. Also, during the project, the economy was starting to subside, capacities were erratic, the company was forced into austerity mode—and ultimately, the plants were closed and the product transferred to other company facilities.

Assembly Process Improvement

The assembly operations consisted of several assembly lines with like product groups produced on the same line when possible. There was an excess of WIP, lines were not balanced, there was no standard work, Takt times were not being followed, the WIP footprint was not managed properly, and there was an absence of a defined and well-managed quality process. There was a high level of dealer and customer product quality complaints. Pre-delivery warranty (the defects a dealer had to address prior to delivering the boat to a customer) was very high. There were a couple of instances where dealers rejected product and sent it back to the plant.

We decided to go line by line with the following process:

Implement 5S

- Conduct a Lean engineering or time study analysis to determine VA/NVA times.
- Determine the critical path time to complete all assembly operations.
- Establish the number of stations by dividing the critical path time by the Takt time.
- Balance the workload by creating and assigning a Takt time cycle's worth of standard work in the form of a job duty for each operator in each station on the line. (See Figures 11.3 and 11.4.)

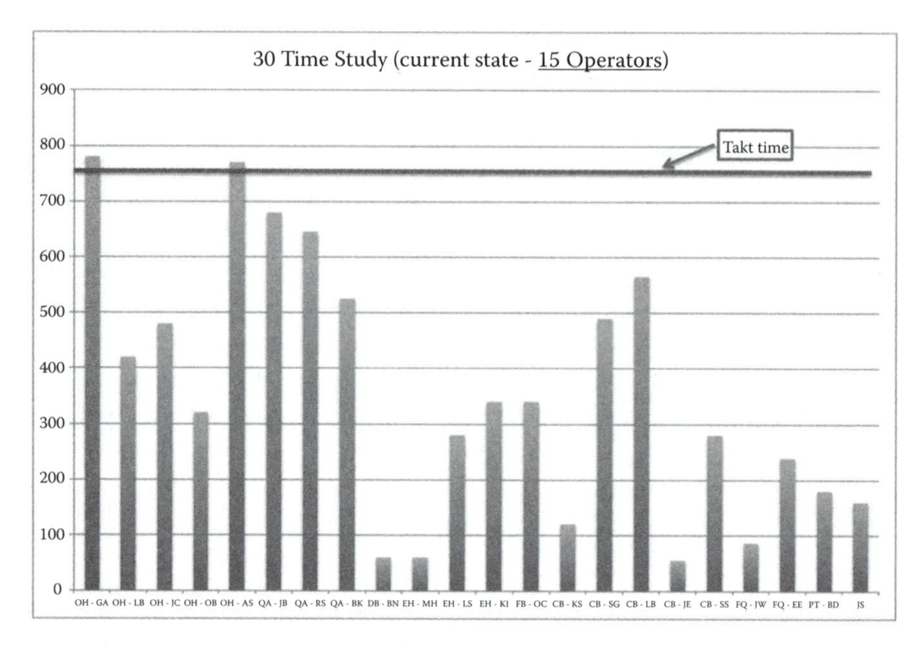

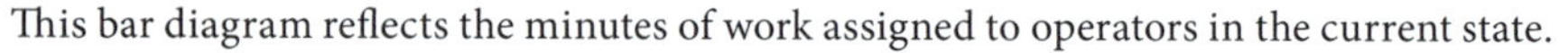

FIGURE 11.3
This bar diagram reflects the minutes of work assigned to operators in the current state.

We set up whiteboards with quality, cost, and schedule data on each line (as previously discussed). There were places to note downtime so root causes could be identified and addressed. Each team met in a huddle at the beginning of each shift and discussed the day's plan, any issues, and what their goals were for the day. Management would review whiteboards throughout the day and respond when necessary with added assistance.

The efforts to implement Lean generated improvement. The total labor cost per unit of the plant was reduced, as reflected in improved labor efficiencies. Note: I understand labor efficiencies are a perilous and frowned-upon metric, but our CFO measured efficiency, and so did we. In this case, however, at the end of the day we were tracking our performance toward producing the same product with less and less labor hours, which was favorably impacted by our improvement activities. The overall materials costs per boat went down with increased disciplines.

Quality Process

Product quality and warranty were significant issues. On the quality front, we developed and implemented a formal quality process. The company

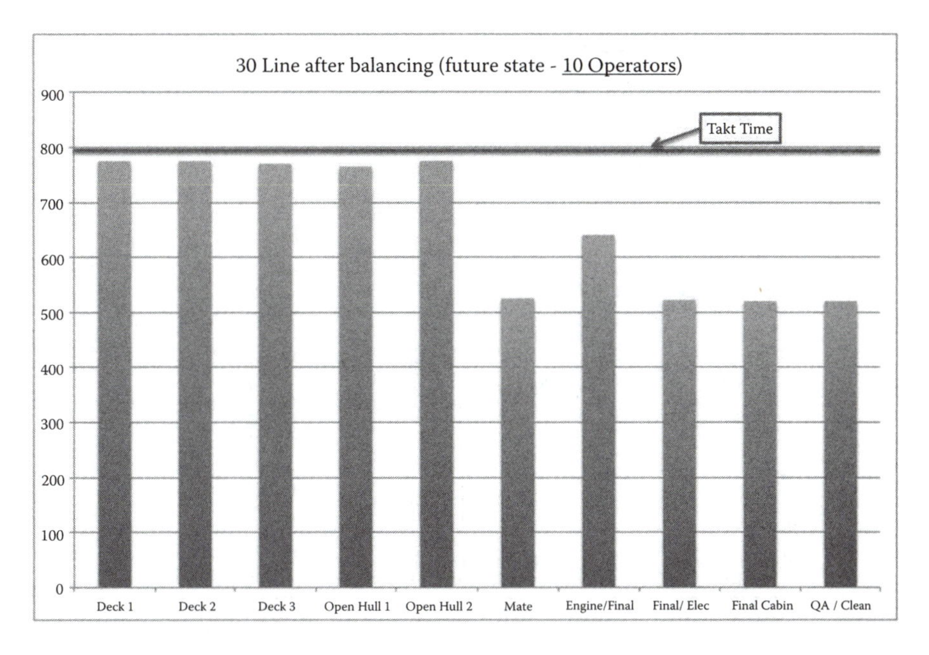

FIGURE 11.4
A bar chart of the minutes of work assigned to operators in process after balancing and improvements.

had high warranty costs, and we needed to work to reduce them. We set up a quality process based upon the critical characteristic approach I had used at Maxum®. If a product or process attribute being wrong caused any of the following conditions, it was considered critical:

- Extreme customer dissatisfaction
- Product or component failure
- A significant downstream production interruption
- An unsafe situation or violation of a regulatory body

We then developed product and process audits. Product audits occurred on every boat. Process audits were accomplished per time period, such as daily, weekly, or monthly. We sometimes used a sampling plan.

We set up a quality laboratory to do incoming materials testing, especially for gelcoat, resin, and adhesives. This was and is very critical. In today's market, chemicals are more and more unstable and vary. It is imperative that materials are tested to avoid big problems.

We focused hard on the lamination process. Equipment maintenance and calibrations were heavily emphasized. Materials shelf life and gelcoat temperatures were monitored, catalyst levels were checked, checks for

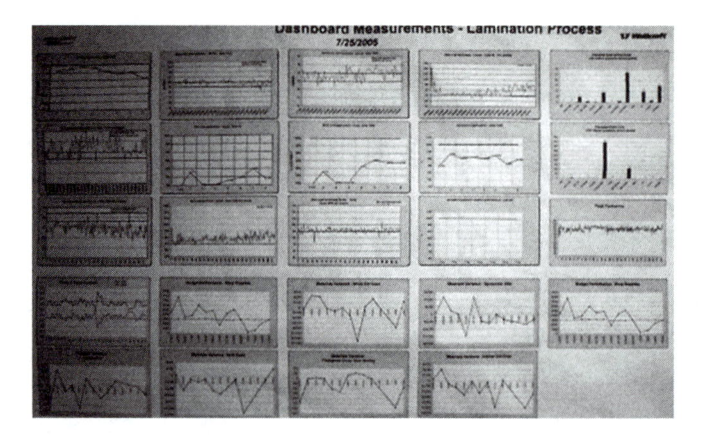

FIGURE 11.5
Dashboard of key measurements of the lamination process.

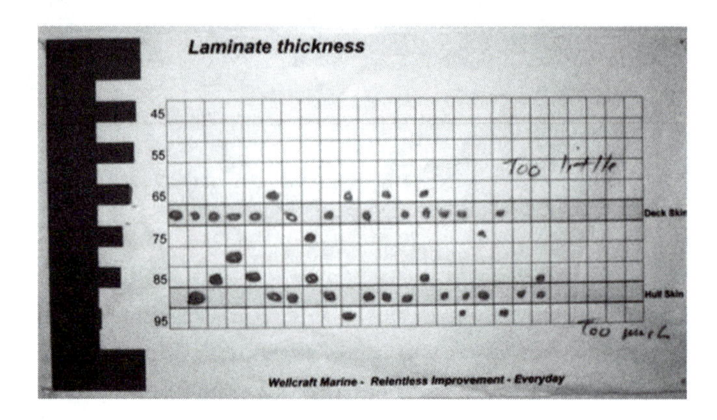

FIGURE 11.6
A control chart created using a depiction of the "mil gage" used by operators to manually verify laminate thickness as they apply it.

peak exotherms being met before applying more fiberglass or resin were done, mil gage use, and materials thicknesses were measured. We then tracked the process measurements and intervened when the metrics indicated a problem. We displayed these process measurements on a dashboard. It was a great tool and stopped a lot of speculation about causes of product issues. It provided great data for us to use in measuring this key process. (See Figures 11.5 and 11.6.)

In the assembly processes, a product audit checklist was done referencing standards: two clamps on fuel lines, proper ventilation, equipment function, proper component installation, etc. were all checked. On an average day, more than 1,000 quality checks were made.

These efforts paid off. We began to get better reports from dealers. Both of our key brands were showing significant improvement, with one earning the most improved in defect reduction in the annual J. D. Power and company surveys, and the other was making progress in customer satisfaction index scores increasing nearly every month.

Lean Six Sigma Cost Reduction Projects

We educated our team leaders and as many employees as we could on the forms of waste and how to identify them, and then we initiated waste reduction projects. A couple of people on the team pursued and earned their Black Belts during the turnaround. We were very fortunate to achieve the level of passion and commitment that we did. It was great. What was even better was that our finance department participated in and supported our cost reduction efforts. Our cost accountant and our CFO regularly attended and participated in project meetings. This was so much appreciated, and their contributions were very valuable. One team saved over $25,000 by switching from bottled water to filtered water, and the water was cleaner! We tested it.

The plant was in Florida and had the biggest light bulb in the world high in the sky over it (the sun). We shut off countless halogen lights in the ceiling and put in skylights to cut thousands of dollars from our electric utility bill.

We set up a rag recycling program that saved us $16,000 per year.

A rubber hose scrap reduction project was initiated, with good results in both the amount of scrap hose being wasted, and the downtime, waiting, and transportation waste resulting from chasing down needed hose.

We focused on mistake-proofing and poka-yoke. Our manufacturing engineers contributed greatly by developing the most effective overlay program I have ever seen. They created fiberglass skins of all decks, liners, and other parts. They then identified on them where all holes should be drilled, and included what holes were for which options. They laminated metal grommets right into them so they were durable. The occurrence of mis-drilled holes decreased significantly. They set up special carts that gave the hole-cutting team quick access to their hole saws, drills, and other information needed to get their job done.

We used dashboards to focus on quality and process performance. I am going to admit something here. OK, here goes. I got way too caught up in charts and graphs. I do think the use of a few key metrics is critical to achieving high performance and problem solving in manufacturing. However, I caution everyone not to get too caught up in charts and graphs. They become sexy after a while, and you can end up spending energy producing high numbers of unneeded information that would be better spent taking action to improve your performance to key metrics. So key measurements are important; just don't fall into "over-processing." I would also like to thank my boss for letting me do this, and knowing I was doing this, letting me learn on my own.

Our team was featured on the cover of *Composites Manufacturing Magazine* for its Lean Six Sigma efforts. There were a couple of great articles inside outlining our approaches and the accomplishments of our team. Our team was recognized by the board of directors as making an important difference in the company, especially during tough times. We all appreciated the recognition. All of us like to be recognized.

Product Development

Like other companies, new product introduction was a challenge; and before going any further, this challenge is not limited to the Sarasota operation I am writing about. Effective new product introduction is an industry-wide problem. Not to say that some companies are not much better than others, because they are. We had a fairly good, defined process from corporate called PLP (product launch process) but engineering would not follow it. They would put on a good front when audited, and then go back to informality. New model introductions were not going well. Engineering was delivering inaccurate build books, incomplete bills of material (BOMs), there were numerous fit issues, and the quality of tooling was poor.

Sarasota actually had what they called a "power team." This was a group of talented boat builders and manufacturing engineers who focused on addressing issues in the first several boats of new model introduction. It was a complete waste of resources, all because we did not set high expectations for our product development process. These boats were built offline. The role of manufacturing engineering was to support production, not finish the prototyping process for engineering. If we can put men on the moon, we can develop a well-designed new boat, accurate BOM, and accurate documentation to have a relatively seamless introduction into the manufacturing process. No excuses.

Why do so many companies expect poor-quality product introductions to yield high-quality product? It is beyond me. Why are there two sets of quality standards? Too often in the marine industry—and I am sure in others—new model introductions consist of inaccurate BOMs, inaccurate or insufficient product documentation, and design flaws. These deficiencies are the input into the manufacturing processes where BOM accuracy and product quality are immediately expected to be of the highest levels. This pays homage to an old comparison between Japanese and American manufacturers. The Japanese were found to take much more time in the product development stage and experienced much more successful product launch stages. Whereas, American companies spend much less time in product development and tend to "throw it over the fence" and let manufacturing and the company's customers work out the details. In my eyes, this phenomenon is a big part of what happened to Outboard Marine Corporation and the Ficht technology.

Everyone was frustrated: the president, myself, our VP of engineering, our brand managers. And, we were getting nowhere. One Saturday, I was working and my boss, the president, happened to be there too. He phoned me in my office just to say hello. He asked, "What are you up to?" I explained to him that I was stressing about new model introduction. production was not getting what they needed from our "model introduction process." There was constant friction between engineering and production, and sales had friction with both engineering and production. Who was it that invented sales volume bonuses anyways? Somebody should slap them. The president asked, "So what do you think we should do?" So I began to rattle off some ideas. I said something like this: "I think we should do …, and I think we should do …, and we shouldn't ever do … unless we do … first." I told him I had it on the board in the office, and he came over to see it. I had drawn a red line on the board. (See Figure 11.7)

The gist of my thoughts was that a new boat should not come across the red line and start in production until a prototype boat was complete and there was a complete and accurate build book, a set of CNC files, a complete BOM, upholstery patterns, and a list of completed jigs and fixtures sitting on the transom of the boat. When engineering had the prototype at this stage, they should schedule a meeting, and the engineering team and the production team would review the prototype boat itself and the documentation placed on the transom. Then, if everything was in order, production would accept the boat into production and begin to build it—and not until.

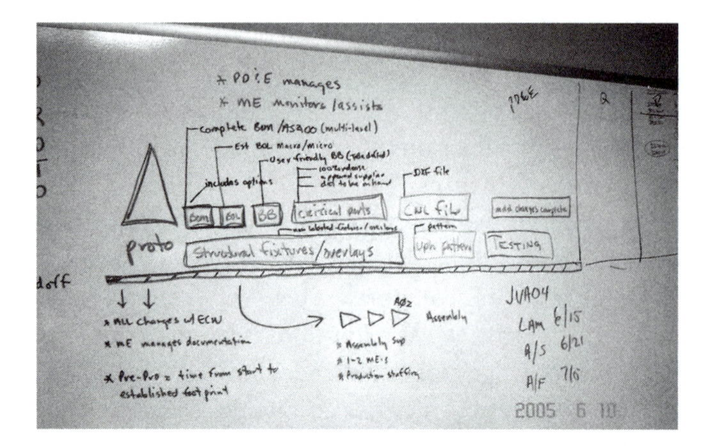

FIGURE 11.7
Rough outline of the deliverables of the new product launch process. The dashed line is the "red line."

The president supported the process, and we began to use it. Following this process gave us much-improved results. We measured the total calendar days between the completion of the first production unit and the eighth production unit of two 30-foot boats. The boat that was introduced meeting the criteria of the "Red Line Transom" process for model introduction reached completion of the eighth unit six times faster—195 days to reach the eighth unit of the boat not using the process, and twenty-four days for the boat that followed the process. (That is 171 days sooner to the eighth unit.) I would say that is significant. I guarantee you that behind every poor new model introduction in just about any industry is a weak development process, or a strong development process that is not followed. (See Figure 11.8.)

There is an old adage to quality at the source: "I will not pass on a defect, nor will I accept one from someone else." Think about it. In a multi-operational machining process where product is progressively machined through several machining centers, operators' defective product would not be allowed to pass from one to the other. Many people get and understand this clearly. But yet, many manufacturing companies in America allow engineering to pass incomplete work product into production.

Lesson Learned

- Quality is paramount in any new model introduction process. Whatever delays are experienced in the development process to get designs and documentation right will be greatly offset by reducing the product launch or introduction phase.

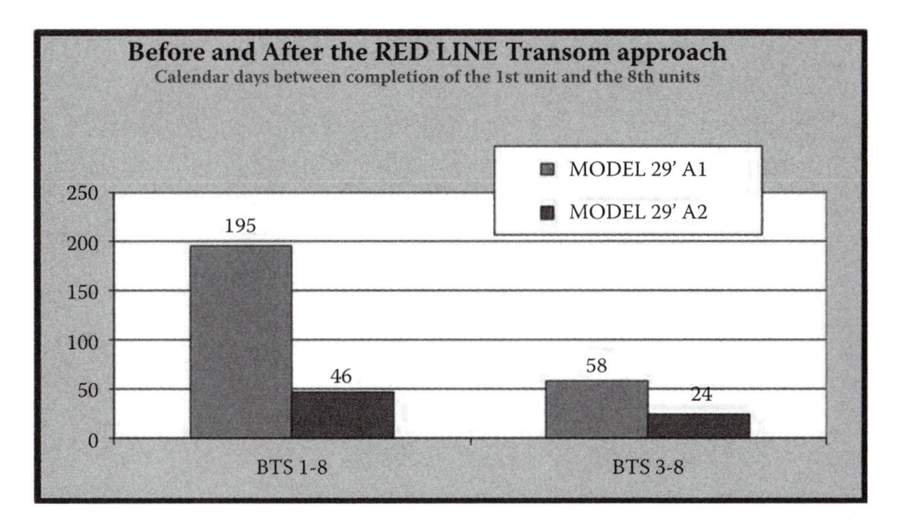

FIGURE 11.8
Comparison of times to complete units with and without a disciplined new product introduction process.

- Setting high standards for the quality of product documentation, design quality, and getting new product right is the most effective way to introduce new product in numerous ways. Nothing that is not right should ever be introduced into manufacturing.

One of the key goals of Lean and Six Sigma is to get things right the first time. Anytime you find a separate area or department set up to address any kind of deficiency or defect, it is not a Lean environment and you can bet there is a good deal of waste present.

12

Improve Design Engineering Process

A few years ago, I was assigned a project to "improve the throughput of a design engineering process for a manufacturing company with multiple plants and products." The company's product was mega-yachts and its focus was on delivering product in a much shorter time than their competitors, ensuring superb quality, and controlling their costs. The company had experienced a high level of growth from a single-plant, entrepreneurial operation to a multi-plant environment. As with any young company, there was a tremendous amount of talented people with high levels of product and process knowledge. However, going from a single- to a multi-plant environment creates a demand for additional skills, and they were struggling with that. The absence of standardization from plant to plant in data, information, approaches, and product was generating a lot of extra work, and a lot of it on the engineering function.

It is not uncommon for it to take more than three years for a customer to take delivery of a mega-yacht, so being able to deliver a mega-yacht in twelve to eighteen months is quite a competitive advantage. However, customers who spend millions of dollars to purchase a product are demanding when it comes to product features and customization. The process of understanding what a customer wants, synthesizing it into design features, and ensuring production capacity and fit within current production and timelines tied to other customer delivery dates is a challenge. Taking it a step further, all changes and agreements must be managed with strict adherence to the sales contract between the customer and the company. This area of the business is tenuous at best; but on the other hand, mastering the ability to quickly design and meet customer desires is literally required in this market. And these goals are right in line with the goals of Lean.

Leading improvement in a high-volume design engineering department is not for the meek. The design engineering (DE) function develops new product; however, the majority of the demand on DE is driven by both customer and internal product change. Generally speaking, the main superstructure from boat to boat within a model stays the same. However, there is a great deal of customer customization in the joinery or interior components, such as cabinetry, wall panels, ceiling arrangement, and a whole host of other areas. Each boat will have a unique lighting plan and usually a customer-selected array of audio-visual and navigational equipment. Rearranging rooms and/or areas on the outside of the yacht provide design challenges for the structural engineering team. On one boat, an owner may want a heli-pad and a fueling station for his helicopter. A submarine is not out of the question. Actually, very little if anything is out of the question, as money is really not a concern for customers who buy mega-yachts. So, if they want it, the engineers can design it.

Engineering was responsible for delivering a plan set to production for the construction of each yacht. The plan set consists of drawings of the standard items that generally do not change from boat to boat, and drawings to support any customer-driven change or any standard drawings requiring modification due to customer change. The average plan set contained more than 1,000 multiple-page drawings. Engineering was also responsible for ensuring compliance with regulations and certain outside certification standards such as the American Bureau of Shipping, or the Cayman Island registry. Design was accomplished primarily in three-dimensional modeling software, with some work being done in AutoCAD and some surface modeling software.

At the start of the project, the stated issues were:

- Documentation provided to production contained excess inaccuracies
- Documentation provided to production was incomplete and insufficient
- Documentation was provided late to production

Initial actions (first week):

- Created a mapping event planning sheet: Excel® spreadsheet with many tabs
- Created a planning calendar
- Set and communicated a date for a process mapping event thirty-five days out
- Arranged place to conduct the mapping event: company aircraft hangar

Pre-event actions (thirty days):

- Developed and distributed an internal customer survey
- Conducted input discussion with key stakeholders: plant managers
- Developed a detailed three-day mapping event agenda
- Began pre-event data collection using checklist
- Selected participants for mapping event and communicated information
- Gave a presentation to the department outlining the improvement process
- Interviewed employees, began relationship building
- Conducted Lean training with department: Waste, Mapping, Flow
- Requested and received a department overview from area managers
- Met with HR to discuss any employee situations necessary
- Started a supply list for mapping event
- Arranged for food and beverage during event
- Walked through and reviewed mapping event site
- Bought and distributed several books on Lean and mapping
- Printed out large waste category diagram for mapping exercise
- Visited production plants: talked with employees, reviewed drawings
- Created training and facilitation presentations for the mapping event
- Printed out the preceding process' map for reference
- Requested overview of strengths and weaknesses of CAD platforms (three)
- Walked the factory floor again and again to understand the manufacturing environment
- Determined the expected outputs from the mapping event:
    - Current state process map
- Future state process map

Engineering/configuration management office layouts:

- Company facility location diagram with engineering staff noted
- Disconnect list
- Undesirable elements list
- An Affinity diagram of engineering issues

A note on all the preparation listed above. In my experience, process or value stream mapping is one of those activities where preparation is "nine-tenths" of the game. First, a good mapping event will tie up valuable people for three to five days. That is a huge cost. Proper and detailed preparation will make sure the time is well used and worth it. If you do not have a good plan, the event can get bumpy. As the leader of the event, you will have a key role in listening and coaching the mapping process. If you are not prepared with a well-laid-out plan, solving logistics issues will take you away from facilitating the event.

Managing projects like this can be very tough and challenging. You will be going into the "valley of change." You are declaring a war on waste. Look at this stage of the project like you have a chance to gather all the intelligence you can on the enemy (i.e., waste) before you get too far out on the battlefield.

The process mapping event was well supported. If you can ever use an aircraft hangar for an event, I strongly recommend it. We taped up four-by-eight-foot sheets of cardboard to the back of the hangar door, which was a garage-type door and only about sixty feet long. We asked a team member to take photographs of the event. I highly recommend this, and use a high-mega-pixel level. This can be a big help in transferring map information to mapping software if desired. We focused a different group of people on each information/value stream, and we had a team create a detailed diagram/map on the relationships between the different IT systems in the company, including those in design engineering.

We started right away by reviewing the three-day agenda, the day's agenda, and the expected outputs of the mapping event. I gave an overview of Lean principles, highlighting the forms of waste, non-value added, and the concept of flow. More training and overview on process mapping techniques, undesirable elements, and disconnects. We then developed an affinity diagram aimed at what was causing insufficient, inaccurate, and late drawings. At the end of each day, we gave a brief of the day's activities to management.

One tool I feel is very valuable is that of a survey. The purpose of the survey was to gain an understanding of how the internal customers viewed the performance of the engineering process or department and how the staff in engineering saw them. I had also received a lot of feedback and comments about engineering from all over the company. That did not mean any of it was true. But I did get the comments. One purpose I had was to make sure that the engineering team understood how the rest of the company saw them. I wanted to make sure that they did not think what was being said was just coming from me. I knew that if they truly understood the situation, they would better support the project.

Conducting the mapping event went quite well. There was a high level of energy, enthusiasm, and participation. There was lots of buzz in the room. So there were four key mapping groups: one for each of the three product information/value streams and one for the IT/configuration management processes. Now mind you, all these people worked in the same basic processes on a daily basis. Literally, every time someone would call out a step in the process, someone else would disagree and suggest that was not the next step or action. Then someone else would jump in. This went on for nearly every yellow sticky note that went on the board. It was very clear that no two people in engineering understood the processes the same way. Wow!

The group spent all of the next day mapping the current state. Once the maps were done, the groups reviewed the maps and annotated on them opportunities to reduce waste or improve flow, or the presence of an undesirable process element. This was a lot of work, and they did a great job. Toward the end of the second day, they started outlining their desired future state. The essence of the future state was a single standardized process and procedures for all value streams in the company.

We wrapped up the event on the third day. We let the group know that we would compile all the information, photos, etc. from the event and place it in a network folder for all to see. We encouraged everyone to communicate back any afterthoughts or ideas they have after they departed.

The initial key projects required to transition to the future state were:

- Define the overall design process and standardize
- Establish a part numbering system
- Establish drawing standards
- Implement a drawing quality process
- Implement a drawing schedule process
- Develop reports and schedules providing information to manage the workload
- Commission a high-level IT project to resolve IT system issues
- Document and improve the change management process

After consolidating the information from the event, we created project teams and reviewed progress weekly to track progress. One thing to remember is to keep the momentum going. People tend to go back to work and have responsibilities, and getting work done toward the future state can be taxing. This is where positive coaching and leadership come in. It can also help to remind everyone of how much better things will be when the future state is achieved.

We were about forty-five days into the project at this point. One of the members of the engineering staff had shown increasing interest in developing the future state and was highly knowledgeable in engineering document control processes. We assigned him to the position of process lead, and he began taking on more and more of a leadership role in driving toward the future state. He and I worked very closely at managing the effort. It was one of the best things I liked about Lean. The combination of my skills in leading change and process improvement, and his significant knowledge of the engineering discipline was powerful. Neither one of us could have been successful without the other, or the efforts of others on the team.

Over the next year we:

- Defined the overall design process: we worked diligently on continuing to develop and update the design engineering process maps. It served as a reference for discussions and problem solving, and was constantly referred to. At all times, the process maps were on the wall and used as reference in discussions.
- Established drawing standards/standard drawing practices: we initiated the use of ANSII (American Standard Code for Information Interchange) drawing standards. This was the proverbial case of trying to fix the bike while riding it. The engineering function maintained thousands of drawings. Hundreds and sometimes over a thousand drawings per month flowed into the production plants, providing documentation for the standard boat and any customer-specific design documentation. Over the course of years, the engineering function veered from using standard ANSII drawing practices or standardization of anything in general. A lack of adequate preparation for the transition to the use of solid modeling, combined with not understanding the value of or implementing disciplined product data management, allowed drawing formats over time to morph to pictures of models absent of the key drawing information, including dimensions. This situation had driven a great deal of angst into the relationship between production and engineering. Working through and resolving this situation was brutal. The tension involved with this situation impeded collaboration in working through it and getting

buy-in to the ANSII format across the company. Ultimately, agreement to move to the standard ANSII drawing format was reached. All future drawings would be created in this format, and current drawings would be converted as resources permitted.

- Established a part numbering system: during this project, a significant opportunity in implementing a proper product data management (PDM) discipline was identified. This required implementing standard industry PDM disciplines in accordance with standard industry engineering document control references. This included proper revision control, use of part configuration, standard naming conventions, and removal of intelligence from part numbers. In addition, metadata was created to aid in reducing data mining times.
- Implemented a drawing quality process: a process was needed to manage the flow of drawing discrepancy feedback from production. There was a lot of it, and it needed to be managed. We established a document quality process that any drawing user could enter discrepancy data via a pdf mark-up tool, a scanned document, or in written form. This process and the redline process were combined. Redlines could be submitted for approval manually, or with a mark-up tool in the change management database. Whenever an engineering staff member worked on a drawing, he or she could identify any discrepancies needing to be addressed prior to releasing the drawing back to production.
- Implemented a drawing scheduling process: engineering's key priority is supplying information to four production facilities at the rate they need it to achieve on-time delivery of the product. The drawing demand is derived from the production schedules of four production facilities. In addition, at any given time, hundreds of document or product changes are in process. Batch processing drawing sets for the mega-yachts was not an option. A system had to be developed to combine, see, and prioritize the drawing demand. A drawing scheduling tool was created by setting up engineering as a work center and routing all make parts through it. The demand for all plants was combined and displayed by order of need. This worked effectively as a priority planning tool.
- Commissioned a high-level IT project to resolve IT system issues: a list of IT issues was developed by the IT function, and a weekly IT conference between the corporate IT team, the engineering IT

team, and management was scheduled. Action items were assigned and followed through until all major issues were resolved. One of the issues identified and resolved in this effort was the need for standard three-dimensional modeling processes.

Future State II: As we drove forward to achieving the future state, more and more disconnects and deficiencies in the current state revealed themselves. It seemed that every time we opened a process door, we found more issues and another door! We continued to update the current state maps and the future state map, and added required projects to achieve the future state. We drove forward.

The absence of an engineering bill of material (BOM) was a significant issue; the engineering function used the production BOMs in the MRP system and had never created or maintained an engineering BOM. The production BOM was formatted to support scheduling production and was not formatted by product structure. Some assembly components were divided among more than one area in the BOM. This made finding data very cumbersome.

The company's change management process was not working well. High levels of change driven by engineering rework and customer-demanded change were occurring. The amount of change was enormous. Customer change was inconsistently managed, and the information provided to engineering to process it was incomplete and insufficiently defined. This situation would be exacerbated during times of slow sales or tough markets, which often delayed the availability of information to engineering to process it as fast as production required to maintain schedule attainment.

The cumulative effect of issues was significant. The engineering function was not keeping up with the demand for information. The absence of product data management, inconsistent modeling practices, and engineering BOMs, and the lack of drawing standards or standard formats over years as the company grew, left a great deal of inconsistency in engineering files and data. Data mining for design information took an exorbitant amount of time, and in many cases a needed file was never located. This impacted the ability of designers to identify all affected drawings or files when making changes. This situation was exacerbated by an increasing level of product change and modification demanded by the customer and driven by insufficient staffing levels in engineering. The factories were starving for information.

The updated future STATE contained the following conditions:

- Effective product data management discipline
- Use of ANSII standard drawing formats and practices
- Improved flow and quality of customer-driven change information to engineering
- An engineering BOM formatted by product structure to reduce data mining time and product interferences driven from design changes in absence of full visibility of impact to other product structure components

The action plan taken to achieve the future state was as follows:

- Finish implementation of product data management software and continued discipline in managing product data
- Creation of a new standard plan set for each model in the ANSII standard formats and practices
- Creation and maintenance of an engineering BOM formatted by product structure
- Increase staffing levels in engineering relative to the workload in supporting production and achieving the future state
- Creation of a manufacturing engineering function in each plant to manage the interface with engineering, coordinate product changes, and provide technical product expertise and assistance to production and engineering

This was a tough project. The new drawing formats were well received by production, engineering began to receive better and more complete customer information, and finding information became easier as all files began to be filed by product structure. Implementing an ANSII drawing format reduced the number of joinery drawings by 30 percent. The manufacturing engineers were well received in the plants and made a huge difference in helping operators keep accomplishing production. During this project, several employees made comments that it was becoming the place they wanted it to be, it was much more organized and less stressful, and that they felt better about where they worked.

Results:

- Increased the average designer drawing throughput by 37 percent
- Reduced the number of drawings required for one of the key drawing categories by 30 percent
- Reduced the number of drawing pages in key category by 41 percent
- Productivity improvements allowed the freeing up of staff to add three manufacturing engineers without adding to total head count

Lessons Learned

- Culture eats strategy for breakfast. Not my saying, but it is true.
- Little menial tasks like managing data, etc., that may not seem important today, over time can create a serious situation.
- Standardized processes and work are critical to success.

13

Improve Materials Management Process

I was requested to lead a project to strengthen and improve a company's materials management process. At the start of the project, the stated issues were:

- Materials stock-outs were disrupting production.
- Excess inventory was tying up working capital.
- The company was receiving a high dollar stream of unplanned invoices for materials ordered outside the system. In one instance, the company reported that it had received over $1 million of unplanned and unexpected invoices after a major project had been completed.
- There were few usable system reports or visibility of information for management.
- There was a companywide consensus that the manufacturing-side ERP (enterprise resource planning) system did not work.

Initial actions (first week):

- Created a mapping event planning sheet
- Created a planning calendar
- Set and communicated a date for a process mapping event
- Arranged place to conduct the mapping event: conference room

Pre-event actions (30 days):

- Interviewed key stakeholders: buyers, materials managers, production leaders, accounting, IS manager, IT staff, executives
- Walked the factory floor and observed materials, inventory, and storage
- Obtained access to ERP system and reviewed system

- Selected and invited event participants
- Sent out a user survey for input regarding the ERP system

Determined the expected outputs from the mapping event:

- Current state process map
- Undesirable elements (UDEs) list
- Disconnect list
- User input and suggested improvements

We conducted the event in a large conference room in one of the manufacturing factories. It is so important to never forget the Gemba. This is where your product or service is created or delivered, and it is the reason the company is there.

The event went well. It always amazes me how enthusiastic and engaged people become in a mapping event. If they have never participated in mapping before, they almost always look a little out of place and confused about how this works. Then, almost every time within thirty minutes of placing sticky notes on a board, they become engrossed and take it away. That was the case in this event. Also, a mapping event is a good place to observe and understand the people you are working with. Informal leaders will emerge, and others display surprising talents, skills, and valuable perspectives. You should keep these folks in mind for future events.

The mapping event yielded a list of over a hundred issues, symptoms, problems, disconnects, UDEs, and a flowchart of the materials processes. Again, there were many disagreements among participants from all across the company regarding how things were done. This is a symptom of a lack of standard work processes.

The key drivers in the current state were:

- The way the company formatted, structured, and entered data into the ERP system would not allow the operations side of the system to function properly and it rendered many key reports unusable.
- There was a lack of clearly defined materials and purchasing procedures.
- No two facilities used the system the same way or managed materials the same way.
- There was an overall lack of knowledge of how to use ERP systems across all levels in the company.

The key elements in the future state were:

- Management of the operations side of the ERP system in accordance with the procedures and approaches outlined in the ERP software user manuals "out of the box."
- Consistently structured BOMs across the company and in a manner that allows the ERP system to function properly.
- A BOM specialist in each factory to maintain the BOMs.
- Standardized materials control procedures across the company
- Standardized purchasing procedures.
- A portfolio of reports of important information to use to run the business.

The key projects required to transition to the future state were:

- Complete analysis of requirements to convert and input data into the system as called for in the system instructions.
- Train a bill of materials specialist position in each factory.
- Develop standard BOM maintenance procedures and apply them consistently.
- Develop a standard bill of materials for each product in the company.
- Develop standard purchasing procedures.
- Create software solution to provide reports.
- Create reports: part shortage report, scheduled item late list, bill of materials, job reports, production schedules, purchasing history report.

All projects to implement the future state were implemented. Here is a summary of improvements gained:

- Significant drop in disruption to the production function due to purchased part stock-outs
- Reduction in stock-outs due to not confirming orders with suppliers
- Elimination of unexpected invoices or charges after product was shipped
- A reduction in inventory from ordering material against requirement generated from inaccurate BOMs
- Reduction in time wasted across the company trying to find out the status of materials availability

14

Lean Transformation Strategies

ELEMENTS OF A SUCCESSFUL LEAN TRANSFORMATION

Develop a Lean Roadmap

This will be a detailed, yet flexible plan to implement Lean across the company. Think about whether you really have the support of upper management. Make sure you are honest with yourself. If the majority of the members of the steering committee are not highly committed, and one of them must be the president, then you will need to approach Lean in a different way. Understanding that there is more than one approach, it should go something like this:

- Get a Lean leader
- Establish a Lean steering committee
- Establish a Lean resource plan—Lean team
- Establish and communicate the Lean Vision to the company
- Develop the roll-out plan
- Implement the plan
- Implement relentless improvement
- Move to advanced Lean practices

Get a Lean Leader

The Lean leader must have a combination of both technical skills with regard to the Lean body of knowledge and people skills. He or she must have solid leadership skills. Implicit in implementing Lean is leading change. This will require skill. It will be challenging, yet rewarding. At times, it may be grueling, yet fulfilling. technical skills alone will not work. Have the Lean leader complete a Lean assessment immediately. There are several online assessments available. Select one and go. It will be a good discussion medium with the Steering Committee.

Establish a Lean Steering Committee

This means management. Get them on board, but remember that their time is limited, so manage their time well. Set up training for the steering committee in the fundamentals of Lean. Tie the Lean strategy into achieving the strategic objectives of the company. Remember: low cost, short cycle times, highest quality, and on-time delivery to customers. In today's business climate, managing working capital and cash flow can be a huge challenge. Lean can be very effective in supporting this effort. Make sure accounting is on the steering committee.

Set up regularly scheduled meetings, at least once per month. Make sure the meetings are well managed. Have an agenda and follow it. Provide updates on Lean activities, progress to the roll-out plan, and then to the Lean roadmap. Develop and distribute a concise report of Lean metrics as you implement them. Always relate the Lean effort to the accomplishment of strategic objectives, which are always at the core of the executive's agenda. Encourage and solicit their participation in small celebration events along the way. It speaks volumes when presidents and executives walk out into the factory and attend.

Have the steering committee participate in some training. Make sure it is good training, and not too long. Also, make sure it is not too technical, but really gets across the principles of Lean and the benefits of a Lean culture—that is short cycle times, lower costs, higher quality, more satisfied customers, a better work environment. Keep them informed and engaged. Invite them to victory or project completion celebrations.

Establish a Lean Resource Plan

One of the key measures of an organization's commitment to Lean is the resources they are willing to deploy in support of it. One of these resources is a Lean leader or champion. Depending on the size of the organization, a Lean team may be in order. I once interviewed a candidate for a Lean leader position. He ended up being more suited for a larger role. However, one thing he did during the interview has stuck with me to this day. He said, "Bill, I would like to talk with the president." I told him I would introduce him. He said, "I'm not interested in meeting him. I want to talk with him. I need to know his resource plan and level of commitment. If he isn't committed, we will be wasting all of our time." He needed a job too. I always admired his wisdom.

Establish a Lean Team in the Company

Lean team members should be trained in Lean and available to support projects and keep them on track. The team can range from one person to many, depending on the size of the company. One person on the team should assist in Lean communications for the company. This can be a Lean newsletter, bulletin board management, training support, process mapping event support, etc.

Establish and Communicate the Lean Vision for the Company

Make sure it includes the what, the why, expectations, and benefits to the company. Make sure to communicate that Lean is *never* about cutting jobs; it is about creating and keeping jobs by allowing the company to sell their product for lower prices, higher quality, and more quickly to customers than to your competitors. Tell them that they are the most valuable asset the company has because of their product and process knowledge, and mean it.

Develop the Roll-Out Plan

This should be a fairly detailed plan, noting that it should be flexible as time goes on. Once again, understanding that there is more than one approach, it should go something like this:

- General training
- 5S implementation
- Map and improve value streams
- Implement Lean metrics and visual management
- Total productive maintenance
- Align product development/new product launch with Lean principles
- Deploy and sustain relentless improvement activities
- Map and improve support processes
- Move to advanced Lean

Implement the Plan

Assign responsibilities, energize, and go. Celebrate often!

Implement Relentless Improvement

Train every possible employee in process improvement skills. The best way to train is to do. Always include training in any improvement activity.

Map and Improve Support Processes

Go through engineering, materials, purchasing, accounting, and sales.

Move to Advanced Lean

Create a new implementation plan and include Hoshin Planning, Hejunka, Lean culture, and values reinforcement.

IMPLEMENT THE ROLL-OUT PLAN

General Training: Lean Values, Overview

Training is important in a Lean culture—not as important as doing, but still important. All leaders should be trained in the fundamentals of Lean. It does not matter if it is on- or off-site, but it needs to be good training. Any leaders desiring more training should be encouraged and supported in doing so. Purchase a Lean library of several books. Have them read *The Goal* (by Eliyahu M. Goldratt), *The Machine That Changed the World* (by Jim Womack, Dan Jones, and Daniel Roos), *Lean Six Sigma That Works* (by Bill Carreira and Bill Trudell), and *Lean Manufacturing That Works* (by Bill Carreira). Emphasize that this is not a program—it is a solid commitment!

5S Implementation

Implementing 5S is challenging, yet it is really rewarding. Done correctly, it will be a great boost to morale. I have never led a 5S implementation that did not strike pride in the employees. One of the best values of 5S at the beginning is that it gives the Lean leader and team a great opportunity to begin to build valuable and important relationships with employees. Do not miss this opportunity to roll up your sleeves and get dirty with your employees. 5S is not housekeeping. It is a philosophy for managing a workplace. Educate, educate, educate. Sort, segregate, shine, standardize, and sustain, sustain, sustain, sustain. Set up a weekly 5S safety audit and score it. Select an audit team member from each area in the plant. Always have them audit an area other than their own. Set a minimum required audit score and give maximum leadership, support, coaching, and focus on any area not meeting it until they do. Celebrate the best scores and keep it positive. Incorporate safety into the audit.

Map and Improve Value Streams

Mapping value streams, creating current state maps, future state maps, and future state transition plans are critical skills and activities. It is imperative that these skills are developed in the Lean team, and the more employees who can learn these skills, the better. Training in Lean principles and 5S should always be part of mapping events, but make sure your training includes a lot of doing. Study up and practice value stream mapping. On one level, it is simple; but on another level, there is a lot to it. Second to being able to map processes is the ability to execute successful Kaizen events. Practice makes perfect.

Implement Lean Metrics and Visual Management

Not having metrics in a Lean environment is like not having a scoreboard at the Super Bowl. If you measure it, they will come. Not only do metrics show performance or indicators, they create focus for the team. If you want the team to focus on something, measure it. Learn to select two or three simple Lean metrics for each area. Only display and use metrics that the employees in that area have the ability to impact and make sure they understand them clearly. Take great care in ensuring that the operators truly understand what the metrics mean and how they are created. Go look at them often. Make sure your president is seen looking at them. That will mean a lot to the employees. If a metric goes in the wrong direction, great! This is an opportunity to use your Lean skills in problem solving. Determine the root cause, use data, and develop a solution. Make it fun and the employees will appreciate the help.

Total Productive Maintenance (TPM)

I will not go into detail here, but TPM focuses on maximizing the up-time of machines and depends heavily on operators to properly maintain their equipment. Machine downtime impacts delivering product on time to customers and drives unnecessary waste into the system. Train a trainer, and then train employees.

Align Product Development/New Product Launch with Lean Principles

Having been involved in many product launches during my career, I cannot overemphasize the amount of waste, frustration, and effort that can be saved by using a disciplined approach to new product development and introduction. It is equally important to teach engineers Lean principles. Do not assume they understand Lean. Design for manufacturability, part standardization, and short product cycle times should be emphasized from the first thought of a design. A company will never have more control over these concepts than they do in the design process. The Japanese learned a long time ago to put a lot of work in up-front. The factory floor is no place to design a product.

Deploy and Sustain Relentless Improvement Activities

Teach problem solving, 5 Why, cause and effect, PDCA (plan, do, check, act), charting, brainstorming, affinity, and mapping skills across the organization. Set and assign goals. Do this until the end of time. Celebrate often!

Map and Improve Support Processes

Go through engineering, materials, purchasing, accounting, and sales.

Kaizen Events

Conducting Kaizen events is an important part of the Lean environment, and Lean implementation as well. Kaizen events allow rapid improvement over a short period of time. Here is my suggestion for relatively simple Kaizen events.

- Select a leader: he or she should be experienced or a professional.
- Select an area or process.
- Select a team.
- Train in the seven forms of waste and 5S, if necessary.
- Send team to area for an hour and have them talk to people, make notes, and document forms of waste. It would be even better to give them a printed audit form with the wastes listed with short definitions and a place to note the instances.
- Have the team create a rough scaled map of the area (include square feet) and equipment layout.
- List all comments from notes on a flipchart.
- Assign values: easy/quick fix or difficult/long term.
- Select the items that will be addressed (top three or four).
- Create a plan to implement.
- Implement the changes; include 5S if not already in place.
- Celebrate!
- Create a weekly checklist to sustain improvements.

Move to Advanced Lean

Create a new implementation plan and include Hoshin Planning, Hejunka, Lean culture, and values reinforcement.

Be a scholar of Lean. Read every Lean book you can. Get as much training as you can. Get as many certifications as you can. Remember that Lean is a journey, not a destination. There is no Lean destination. There is only the journey. Use your knowledge to teach your people

Educate everyone, and I mean everyone (including the president), in the forms of waste and the concepts of flowing product through the value stream. Display simple diagrams of Lean concepts in places employees will see and read them.

Become a scholar of change management. Learn about people and how they react to change. Communicate all change with the change mantra. "This is what we are doing, this is why we are doing it, this is what is expected of you, and this is what it does for you. Always, always assume someone with a bad attitude is just a human being struggling with change, and do everything you can to help them through it. My experience has taught me that there are a lot more people struggling with change than there are people with bad attitudes. Remember that someone who cannot read will often act out in a uniquely out-of-character way. If you have an employee acting totally out of character during change, consider whether or not they are able to read. This situation can be dealt with easily, and you can help them, once you know it. Be sensitive.

MISCELLANEOUS TOPICS

Communicate in groups. Talk one on one. Give people books or pamphlets on Lean concepts. Remind them how important they are. Go in the factory. Lots. If you are not going through four pairs of work boots a year, you might not be where you should be. Be the ambassador of Lean. Always get back to employees, and keep every commitment you make.

Get HR involved. Teach your leaders to build relationships with HR. HR roles differ greatly from company to company. Try to involve them in all Lean activities and training. Encourage them to update themselves on Lean and change management. They can be a valuable asset in helping people through the change.

Understand the "valley of change". One pattern I have seen over and over in my experience is the "valley of change." Change is kind of like a relationship. When the idea is new, it is exciting, it is no problem. That is because talking about it is easy. Then the work starts. The current state is still going pretty good. You have developed the future state and begin to implement. There are lots of projects. People have work to do, and even good change is stressful. Now people are actually being asked to do things differently, and humans struggle with "differently." Going Lean can be counter-intuitive. It goes against traditional approaches. So employees are struggling with the change, and others in management are uneasy with it. You have entered the valley of change and it can be a tough place. You must follow your process and drive toward the future state, like following a compass in the forest. If your compass says go straight no matter what, then go straight even if the place to the left looks like where you ought to be going. It is going to get worse before it gets better. It always does. It may feel like you are going to fail badly three weeks before you succeed greatly. It is like losing weight. You diet and diet and diet, and you do not lose. You have been doing this for weeks. You are getting frustrated. Then one day someone says, "Hey, you look like you're losing weight," and you go home and weigh yourself, and you have lost five pounds! You have to stick with it. Do not give up. Improvement will happen.

Be patient with presidents and CEOs. Their job in today's economy is akin to riding a bull. When you shake his hand, you are only seeing the tip of the iceberg. Every president is dealing with ten ugly things that no one but he even knows about. Your relationship with the president is going to have a big effect on what you get done.

Do not call it Lean. If you are in a company that is not buying into Lean, try not calling it Lean. In some situations, using Lean terms might confuse people or turn them off—such as Kaizen, Hejunka, poke-yoke, Hoshin planning. Know what I mean? Instead of asking to do a Kaizen event, ask to get a few employees together and straighten out an area. Instead of talking about 5S, just say, "We want to clean up the place." Instead of Hejunka, just say, "We want to level the workload." Then after some successes, ease in the Lean vernacular. Hey, whatever it takes, right?

It is not just about technology. It is about culture, leadership, a value system, and respect. The principles of the Toyota production system far outweigh any technology that might be purchased. Some or many engineering schools are adopting Lean into their curriculums. Why? Think about it. How many engineering consulting firms do you see? How many engineering books are on Amazon.com compared to books on Lean or Six Sigma? Too many engineers are arrogant and consider themselves above most others. They are like "ring knockers" from West Point in the Army. They consider themselves better because they have a technical education that is tough to earn. They consider their knowledge the most valuable in the company, even if they do not admit it. However, this flies in the face of the Lean culture, where those with the least education are considered most valuable! Wow, that is hard to swallow. isn't it? But it is also true. It is the knowledge of operators that is considered most valuable in a Lean culture. Their knowledge is key to continuous improvement and is the foundation to all improvement activities.

The attitude toward the operators in the Lean culture somewhat mirrors that of the U.S. Military toward their soldiers. I served in the U.S. Army for seven years, three years as an enlisted soldier and four years as an officer. During both hitches, it was emphasized again and again that everyone in the military was there to support the 11B Infantry soldier or "11 Bush" or "ground pounder." It was all about making sure the soldier had the necessary training, weapons, equipment, food, water, and ammunition to win on the battlefield. These soldiers were on the "FEBA," or forward edge of the battlefield. This was where the action was! There is also a huge effort to make sure the soldier's family is taken care of, and that he or she knows it. The military leadership wants the soldier to be focusing on killing the enemy, not wondering whether his family is safe, or whether their needs are being taken care of.

15

Various

SURVIVING

"Surviving" – I think one of the problems with the American manufacturing culture—and it may just be the American culture—is that company presidents are more focused on their personal survival than on the company's survival. I will categorize this as "it just is what it is." This comment is not meant to be mean spirited, but an articulation of what I see as an issue. Actions that are required for a company president to survive versus what it might take for the business to survive in the long term may be very different. Companies grow and change like people. They go from the entrepreneurial stage, to the high growth stage, to the large privately held stage, and often and finally to the publicly held stage. A president who has presided over the entrepreneurial and high growth stage may not have the skills or knowledge to provide the leadership and actions the company needs at the large private company stage. For example, a company that has gone from an entrepreneurial company to a large company will most likely require more complex systems and strategies. The amount of information and daily activities multiplies geometrically with the growth. If the president does not have the background to provide the needed leadership, knowledge, or the confidence to hire someone who does, the company can stagnate for a long period of time. The company can enter a stage where it is profitable and highly inefficient at the same time as an entrepreneurial company early in growth most likely has generous profit margins due to fewer competitors. Other significant factors come into play. The company may have grown so much that it has outgrown the technical skills of the very people who helped make the entrepreneurial venture a success. Bringing in talent from outside the company and putting them in key positions, some supervising people who have been with the company for years, can create tough challenges, loyalties, and relationships. It takes real leadership courage to make these decisions, and sometimes it takes a new leader.

ORGANIZATIONAL POLITICS

Organizational politics, the tenth waste. Maybe the reason I have so much disdain for politics is that I have virtually no political skills. I am not so naïve as to not understand that politics are a way of life. However, I do not like politics. They get in the way of implementing Lean Six Sigma and they slow down improvement.

I have read countless Lean books, manufacturing excellence related books, Six Sigma books, and I have never seen an article on how to be a good politician. Politics are a huge part of American business, and I see it as a real problem. Google® the term "office politics" and you will find quite a list of books about office politics in America. Google the term "office politics in Japan" and you will not find much. One of the biggest leadership failures of leaders is to tolerate or participate in office politics.

The primary issue I have with organizational politics is that all politics come at the expense of teamwork. Teamwork is about putting the goals and needs of the team first, and the team should make decisions regarding or impacting the team. Back-office discussions with certain members or "sub-teams" undermine the success of THE team.

The primary goal of a political person is to move their personal agenda forward. I have seen people who are literally professional politicians. In Lean, our tools are process mapping, brainstorming, fishbone analysis, Kaizen events, work cell layout, waste identification and elimination, affinity diagrams, spaghetti charts, leveling, line balancing, Hejunka, etc. The tools of a politician are sneakiness, stealth, collusion, alliance building, never giving away your position, giving evasive or unclear answers, holding back information, personal attack, drama, and on and on. If a person in a leadership position is a politician, his people fall right in line and tend to become political too. Not good.

Political skills and behaviors fly right in the face of effective problem solving. In problem solving, it is helpful to know what people really think. It is helpful to take positions on problems, work as a team with everyone, make sure everything everyone knows about a problem situation is out there, and to make solving the problem a higher priority than one's personal agenda.

Politics are about "positioning" issues or views in people's minds. They are about persuasion, innuendo, and power. Facts and data are only used if they support the politician's role, and they hate data. When confronted with a problem, a politician will meet with his or her allies, or schedule a showdown. A leader will get the process participants together and apply problem-solving tools to identify and solve the root cause.

In my experience, people literally have convulsions over data refuting their keenly concocted political position. It is almost funny. I have often said that if you want to see the politicians in the room, throw some data on the table and just watch. Where did this come from? How do we know this is right? This has nothing to do with what we are talking about!

Now, a problem solver would more likely respond with, "Great, we have some data. How'd you get it? Can you take us through it and help us understand?" Their only goal is to solve the problem. They are more interested in knowing the specifics and determining the root cause, not political positioning.

I worked with a junior plant manager on one occasion. He had some issues with scrap rates at the CNC (computer numerical control) router. The president's assistant called me and said that the plant manager had requested a meeting with him, me, and the president to discuss CNC issues. I am thinking, why would he do that? He had not even talked to me! We have a CNC process management team that is tracking all pertinent data, and continually applying a Pareto analysis and driving down issues. Actually, they were doing a great job. The plant manager was even on the team! I asked the president's assistant to cancel the meeting and scheduled an immediate meeting with the CNC process team. The plant manager never showed.

Politics has no place in manufacturing and should be eliminated at every opportunity. It wastes time and is often unethical and disruptive. Sometimes, very talented people leave organizations because of politics.

LEADERSHIP

Implementing Lean takes leadership. You cannot have great implementation with weak leadership. Lean is about change, and teaching people new approaches, philosophies, and values.

The art of leadership can be an admirable thing. Watching a great leader in action is enjoyable and rewarding. Leadership is absolutely a journey. My personal belief is that leaders, with time and effort, can continually hone their leadership skills and become better leaders. However, in my experience, there are some very good people who will never be good leaders. It takes some natural ability, ability that can be honed, but it cannot be created.

More complex areas or challenges may require more time and detail. It is important to empower the team and let them know what their limits are—and the fewer, the better.

HOW TO CONDUCT A PROCESS MAPPING EVENT

It is best if you can take thirty days to prepare for an event. If you can't, you can't, but the more prepared you are, the greater success you will have.

Complete a SIPOC (suppliers, inputs, process, outputs, and customers) diagram. Right up front, it is a good idea to understand who and what the suppliers, inputs, processes, outputs, and customers are.

Set up a pre-event checklist. You will most likely be committing some people to the event, and good planning will make sure you use their valuable time effectively.

Set up an itinerary for the mapping event. Most likely, you are going to want a short presentation on the fundamentals of Lean, some clear definitions of waste, and the fundamentals of process mapping, including symbols, etc.

Have some slides and info on spaghetti diagrams, work sampling, and space utilization analysis, and affinity diagram basics. As you go through each one of those during the event, you can put them up on the overhead for reference.

It is not a bad idea to squeeze a quick mapping training session in prior to the event and then review again in the event.

I recommend sending out a survey across a good cross-section of the areas of the organization that interact on any level with the process(es)s that are being mapped. Ask general yes-and-no and open-ended questions.

You must be Columbo. I do not think you can possibly have too much information going into a mapping event.

Select the individuals for the event and contact them or their supervisor to get approval. You want "doers," people who know the process, who work in the process. You are not looking for people who "know about the process"; you are looking for people who KNOW the process.

Make sure you designate someone to take tons of pictures with a high-mega-pixel camera so you can blow them up to put in mapping software and a million other uses down the road.

I generally follow this itinerary when conducting an event:

DAY 1

- Review concepts of wastes, non-value added, and value streams
- Walk about the areas being mapped
- Discussion - undesirable process elements
- Root Cause / brainstorm / cause and effect
- Map the current state / gather work sampling during session

DAY 2

- Complete mapping the current state
- Review Takt times, value / non-value added
- Evaluate work sampling data
- Begin mapping future state

DAY 3

- Complete mapping future state
- Develop list of projects required to achieve future state

I would use caution in taking away from it, but be liberal in what you add. Lean is an art, so do not be afraid to adjust as necessary. Just remember that time management is an important part of an event. You have a lot to get done, and it often takes more time than expected. By the way, if it can be done, I recommend taking up to five days to do the baseline event. Sometimes eyebrows get raised at that. This flies in the face of reality, does it not? To me, five days of five to seven people's time is a pretty fair exchange for solving a major problem area's issues. Think of all the wasted time the issues in the problem area are driving now!

Note the events and actions in the itinerary. You should have PowerPoint® slides to quickly review the basics for each of these events. The Lean concepts presentation on the fundamentals of Lean on Day 1 will have the most slides. But remember: events are about *doing*, so keep the presentations short.

At the end of each day, management should be brought in (and the more, the better) for a short, stand-up review of the day's activities. This is the time for the members of the group to shine. As the facilitator, your job is to put them in the limelight and show off their work. This can be a very useful thirty minutes, as it will be hard for management not to appreciate the enthusiasm in the team members as they go through their data.

Suggested outputs of the current state are:

- The current state map
- A spaghetti diagram with travel distances and depictions
- Value-/non-value-added percentage analysis
- Activity category percentages
- A space utilization analysis
- A list of undesirable elements of the process (UDEs)
- An Affinity diagram

The current state map should have opportunities to improve flow identified on it, along with process disconnects (a list can also be done).

All this information will be used to create a future state map. The team should carefully review all the outputs together and brainstorm to develop the future state. Create the best-case scenario absent as many of the issues identified in the current state as possible. This should be depicted in map form.

Once the future state map is done, a project "shake-out" list should be created. This is a detailed list of all projects necessary to transition from the current state to the future state. Once that is complete, a priority matrix analysis should be done to identify the easiest and quickest projects with the biggest paybacks to be executed first. Caution should be used in not taking on "save the world projects" that take forever to accomplish. A project plan that gets the best overall return on effort over the most reasonable time frame should be taken.

Once the future state map and the project action plan are complete, the team should review it with management. I recommend taking one to two weeks to give management time to digest and approve. You, as the project leader, can refine the plan during this time. I highly recommend having a one-on-one with the CFO about the effort. Having the CFO in your corner is always a good thing. Make sure he or she is in agreement with the direction and objectives. It is well worth the effort. It is important to keep the team informed on intended actions. If too much time goes by without any action, the energy will go from excitement to disillusion quickly.

About the Author

Bill Trudell is a passionate manufacturing professional and scholar of Lean Six Sigma. He has worked in manufacturing for twenty-five years in positions ranging from assembly line worker to middle management roles to vice president of quality and process improvement. He is currently president of Relentless Excellence, LLC, a Lean Six Sigma practice, and has led or been directly involved in Lean Six Sigma transformations and projects for more than twenty years leading three factory Lean Six Sigma transformations. Mr. Trudell graduated from the University of Tennessee and earned an MBA from Jacksonville State University. He holds the ASQ Certified Six Sigma Black Belt and the University of Michigan Lean Certificate. He is APIC and NAPM certified and has attended training at the University of Virginia's Darden School of Business and Northwestern University's Kellogg Graduate School of Management. Mr. Trudell co-authored *Lean Six Sigma That Works* published by AMACOM and his efforts were highlighted on the April 2007 cover of *Composites Manufacturing Magazine*. He currently offers Lean Six Sigma leadership services through http://www.RelentlessExcellence.com.

Index